Chemistry Research and Applications

CALIXARENE COMPLEXES WITH SOLVENT MOLECULES

CHEMISTRY RESEARCH AND APPLICATIONS

Handbook on Mass Spectrometry: Instrumentation, Data and Analysis, and Applications
J. K. Lang (Editor)
2009. 978-1-60741-580-0

Applied Electrochemistry
Vijay G. Singh (Editor)
2009. 978-1-60876-208-8

Energetic Materials: Chemistry, Hazards and Environmental Aspects
Jake R. Howell and Timothy E. Fletcher (Editors)
2010. 978-1-60876-267-5

Chemical Crystallography
Bryan L. Connelly (Editor)
2010. 978-1-60876-281-1

Handbook of Inorganic Chemistry Research
Desiree A. Morrison (Editor)
2010. 978-1-61668-010-7

Solid State Electrochemistry
Thomas G. Willard (Editor)
2010. 978-1-60876-429-7

Rock Chemistry
Basilio Macías and Fidel Guajardo (Editors)
2010. 978-1-60876-563-8

Electrolysis: Theory, Types and Applications
Shing Kuai and Ji Meng (Editors)
2010. 978-1-60876-619-2

Macrocyclic Chemistry: New Research Developments
Dániel W. Fitzpatrick and Henry J. Ulrich (Editors)
2010. 978-1-60876-896-7

Mathematical Chemistry
W. I. Hong (Editor)
2010. 978-1-60876-894-3

Wet Electrochemical Detection of Organic Impurities
*F. Manea, C. Radovan, S. Picken
and J. Schoonman (Authors)*
2010. 978-1-61668-661-1

Physical Organic Chemistry: New Developments
Karl T. Burley (Editor)
2010. 978-1-61668-435-8

**Chemical Sensors: Properties, Performance
and Applications**
Ronald V. Harrison (Editor)
2010. 978-1-60741-897-9

**Heterocyclic Compounds: Synthesis, Properties
and Applications**
Kristian Nylund and Peder Johansson (Editors)
2010. 978-1-60876-368-9

Influence of the Solvents on Some Radical Reactions
*Gennady E. Zaikov, Roman G. Makitra,
Galina G. Midyana and Liliya I. Bazylyak (Editors)*
2010. 978-1-60876-635-2

**Dzhemilev Reaction in Organic and
Organometallic Synthesis**
Vladimir A.D'yakonov (Author)
2010. 978-1-60876-683-3

**Analytical Chemistry of Cadmium:
Sample Pre-Treatment and Determination Methods**
Antonio Moreda-Piñeiro and Jorge Moreda-Piñeiro (Authors)
2010. 978-1-60876-808-0

**Binary Aqueous and CO_2 Containing
Mixtures and the Krichevskii Parameter**
*Aziz I. Abdulagatov, Ilmutdin M. Abdulagatov
and Gennadii V. Stepanov (Authors)*
2010. 978-1-60876-990-2

Advances in Adsorption Technology
Bidyut Baran Saha and Kim Choon Ng (Editors)
2010. 978-1-60876-833-2

Molecular Symmetry and Fuzzy Symmetry
Xuezhuang Zhao (Authors)
2010. 978-1-61668-528-7

Quantum Frontiers of Atoms and Molecules
Mihai V. Putz (Editor)
2010. 978-1-61668-158-6

**Modification and Preparation of Membrane
in Supercritical Carbon Dioxide**
*Guang-Ming Qiu, Rui Tian, Yang Qiu
and You-Yi Xu (Authors)*
2010. 978-1-60876-905-6

**Thermostable Polycyanurates: Synthesis, Modification,
Structure and Properties**
Alexander Fainleib (Editor)
2010. 978-1-60876-907-0

Combustion Synthesis of Advanced Materials
B. B. Khina (Author)
2010. 978-1-60876-977-3

**Structure and Properties of Particulate-Filled
Polymer Composites: The Fractal Analysis**
G. V. Kozlov, Y. G. Yanovskii and G. E. Zaikov (Authors)
2010. 978-1-60876-999-5

Information Origins of the Chemical Bond
Roman F. Nalewajski (Author)
2010. 978-1-61668-305-4

Cyclic B-Ketoesters: Synthesis and Reactions
M.A. Metwally and E. G. Sadek (Authors)
2010. 978-1-61668-282-8

Boron Hydrides, High Potential Hydrogen Storage Materials
Umit B. Demirci and Philippe Miele (Editors)
2010. 978-1-61668-361-0

Electrochemical Oxidation and Corrosion of Metals
Elena P. Grishina and Andrew V. Noskov (Authors)
2010. 978-1-61668-329-0

**Tetraazacyclotetradecane Species as Models
of the Polyazacrown Macrocycles**
Ryszard B. Nazarski (Author)
2010. 978-1-61668-487-7

**Chemical Reactions in Gas, Liquid and Solid Phases:
Synthesis, Properties and Application**
G.E. Zaikov and R.M. Kozlowski (Editors)
2010. 978-1-61668-671-0

Ion Transfer at Liquid/Liquid Interfaces
Rodrigo Alejandro Iglesias and Sergio Alberto Dassie (Authors)
2010. 978-1-61668-684-0

Molybdenum and Tungsten Cofactor Model Chemistry
Carola Schulzke and Prinson P. Samuel (Authors)
2010. 978-1-61668-750-2

Calixarene Complexes with Solvent Molecules
O.V. Surov, M.I. Voronova and A.G. Zakharov (Authors)
2010. 978-1-61668-755-7

**Nobel Laureates and Nanotechnologies
of Applied Quantum Chemistry**
Vladimir A. Babkin and Gennady E. Zaikov (Editors)
2010. 978-1-61668-849-3

Lanthanum: Compounds, Production and Applications
Ryan J. Moore (Editor)
2010. 978-1-61728-111-2

**Spin-Orbit Interactions in PtF6 and in Related
Octahedral Molecules and Fluorocomplexes**
Svetlana G. Kozlova and Svyatoslav P. Gabuda (Authors)
2010. 978-1-61668-838-7

Interactions of Aqueous-Organic Mixtures with Cellulose
M.I. Voronova, O.V. Surov and A.G. Zakharov (Authors)
2010. 978-1-61668-766-3

Chemistry Research and Applications

CALIXARENE COMPLEXES WITH SOLVENT MOLECULES

O.V. SUROV
M.I. VORONOVA
AND
A.G. ZAKHAROV

Nova Science Publishers, Inc.
New York

Copyright © 2010 by Nova Science Publishers, Inc.

All rights reserved. No part of this book may be reproduced, stored in a retrieval system or transmitted in any form or by any means: electronic, electrostatic, magnetic, tape, mechanical photocopying, recording or otherwise without the written permission of the Publisher.

For permission to use material from this book please contact us:
Telephone 631-231-7269; Fax 631-231-8175
Web Site: http://www.novapublishers.com

NOTICE TO THE READER

The Publisher has taken reasonable care in the preparation of this book, but makes no expressed or implied warranty of any kind and assumes no responsibility for any errors or omissions. No liability is assumed for incidental or consequential damages in connection with or arising out of information contained in this book. The Publisher shall not be liable for any special, consequential, or exemplary damages resulting, in whole or in part, from the readers' use of, or reliance upon, this material.

Independent verification should be sought for any data, advice or recommendations contained in this book. In addition, no responsibility is assumed by the publisher for any injury and/or damage to persons or property arising from any methods, products, instructions, ideas or otherwise contained in this publication.

This publication is designed to provide accurate and authoritative information with regard to the subject matter covered herein. It is sold with the clear understanding that the Publisher is not engaged in rendering legal or any other professional services. If legal or any other expert assistance is required, the services of a competent person should be sought. FROM A DECLARATION OF PARTICIPANTS JOINTLY ADOPTED BY A COMMITTEE OF THE AMERICAN BAR ASSOCIATION AND A COMMITTEE OF PUBLISHERS.

Library of Congress Cataloging-in-Publication Data

Surov, O. V.
  Calixarene complexes with solvent molecules / O.V. Surov, M.I. Voronova, A.G. Zakharov.
    p. cm.
  Includes index.
  ISBN 978-1-61668-755-7 (softcover)
  1. Calixarenes--Solubility. I. Zakharov, A. G. II. Voronova, M. I. III. Title.
  QD341.P5S87 2009
  547'.7--dc22
                          2010013749

Published by Nova Science Publishers, Inc. ✣ *New York*

CONTENTS

Preface		xi
Chapter 1	Introduction	1
Chapter 2	Concurrent Insertion of Solvent Molecules in Solutions	7
Chapter 3	Gas-Phase Interactions of Calixarenes with Molecular Guests	11
Chapter 4	Adsorption of Organic Vapors within Calixarene Thin Films	27
Chapter 5	Molecular Recognition of Organic Guest Vapors by Solid Calixarenes	31
Chapter 6	Transport of Small Mobile Particles through a Calixarene Crystal	39
References		45
Index		57

PREFACE

Solid calixarenes as receptors are characterized by phase transitions, which occur in binding of neutral molecules of guest compounds with the formation of clathrates or guest–host intercalation compounds. The binding of a guest by a solid host compound is largely a cooperative process. It begins after the attainment of a certain threshold thermodynamic activity (relative pressure p/p_0) of the guest and is completed in a narrow range of p/p_0 values. The process of clathrate formation in the binding of a guest by a host is accompanied by a significant rearrangement of the packing of host molecules, and the result of this rearrangement can depend considerably on the molecular structure of the guest. Moreover, the "induced correspondence" of the clathrate structure to the molecular structure of the guest is observed; in some cases, this effect endows solid hosts with improved selectivity.

Scattered information is available as to guest-removal, -addition, and – exchange properties of many host systems, including calix[n]arene derivatives. Upon recrystallization, host can form stoichiometric guest-host adducts, where guest molecules are included in each cavity maintained by the hydrogen-bonded network of the host. Volatile guests can be removed upon heating, to give polycrystalline guest-free apohost. The apohost subsequently binds various guests not only as liquids or gases but also, in some cases, as solids in the same guest-host stoichiometry as in recrystallization. The resulting adducts exhibit the same X-ray powder diffractions as the corresponding single-crystalline samples. Guest exchange also occurs. Despite such phenomenologically rich information, we are still far away from a thorough understanding of how solid-state complexation takes place. There are number of fundamental questions which are related to each other: Do the solid host and its solid adduct share the same phase or constitute different phases? Do guest-binding cavities maintained by an organic network survive or collapse upon guest- removal? How do guest molecules diffuse in the solid host?

Chapter 1

INTRODUCTION

Calix[n]arenes represent an interesting class of preorganized aromatic hosts exhibiting an enhanced ability for cation-π interaction and inclusion of small neutral organic guests. In many biological systems, metal cation-π interactions play an important role in molecular recognition, and atmospheric detection of odorant vapours of organic compounds is one of the most important problems of environment monitoring. Moreover the recognition of neutral organic molecules and cations by synthetic receptors is a topic of current interest in supramolecular and analytical chemistry. It was shown that some compounds such crown ethers and calixarenes forming inclusion complexes with some organic guest molecules and cations can be used for the development of sensors and components of microelectronic systems [1]. The growing interest in these materials is due to the simplicity of their synthesis, thermal stability and the extreme ease of deposition under thin film form [2-5]. From the study of crystal structures of calixarene hosts including organic molecules and research of host-guest calixarene chemistry in the gas phase by mass spectroscopy as well as results obtained by solution chemistry, several conclusions of general validity were drawn [6]. Probably, the strength of the host-guest interaction depends on the potential guest molecule nature, the cavity size and the conformation of the macrocycles, substituents on the upper and lower rim of the calixarenes which influence the cavity size, the conformation and the flexibility of the host molecule, the number of the ligating sites of the host (e.g., the number of the oxygen atoms in the polyether ring). In spite of available facts, conclusions regarding the relationship between the structure and the binding properties of a particular host are rather difficult because of the complex relations between the complex formation constants and the structure of both the host and the guest molecules.

There is evidence that the small neutral guest selectivity in the cavity of solid calixarene host is closely related to the free energy of complexation in solution [7]. In the last ten years, host-guest chemistry in the gas phase has been studied by mass spectroscopy [8], and it turns out that the ionization mode determines whether the results obtained by mass spectroscopy reflect those of solution chemistry. Shinkai *et al.* [9] have studied organic cation complexes with several calix[n]arenes of differing conformation and ring size by the use of positive secondary ion mass spectrometry. Relative peak intensities have been shown to reflect the complex stability in the gas phase but selectivity of the complexation with respect to the size of both the host and the guest differs greatly from that observed in solution. Whereas the conformation selectivity found in the gas phase paralleled that of the solution, the hole-size selectivity of guests was found to be different for the gas and the condensed phases. Despite the enormous work developed to characterize calixarene receptors and investigate their chemistry in solution, very few studies have been undertaken to date to investigate the interaction between them and organic guest molecules in the gas phase [10]. However gas-phase studies provide interesting perspective for host-guest interactions. As the solvent is absent, no solvation effects can modify the electronic and thermodynamic properties nor the geometrical constraints of supramolecular binding, so that pure intrinsic interaction between the two counterparts is uncovered, free from any third-body influence.

There has been much recent interest in organic and metal-organic network materials whose guest-binding properties are reminiscent of traditional zeolites. Lattice inclusion compounds have so far been studied mostly from the static viewpoint (stoichiometry, crystal structure, selectivity, etc.) on recrystallized host-guest adducts. In the context of zeolite analogues, it is also essential to know how preformed solid hosts interact with guests.

As was shown in a study of polymorphism by Atwood et al. [11, 12], under certain conditions, the wellknown compound-host, *p*-tert-butylcalix[4]arene, underwent a phase transitions in single crystals. The authors showed that *p*-tert-butylcalix[4]arene could form a bilayer-type structure, where calixarenes are slightly shifted relative to each other. This results in the formation of isolated cavities with approximate volume 235 $Å^3$ per dimeric unit. In spite of the inaccessibility of these cavities, the crystal is capable of adsorption of guest molecules into its lattice with a side shift of adjacent bilayers relative to each other. The surface of bilayers is formed by bulky *tert*-butyl groups alternating with calixarene cavities and gaps between neighboring *tert*-butyl fragments. It is therefore easy to image how neighboring layers can slip one above other favoring the absorption or removal of guests in the intercalation reaction of the solid–liquid or solid–vapor type.

The temperature dependences of vapor pressure of some calix[4]arenes and their complexes with solvent molecules were determined by the Knudsen effusion method [13, 14]. It was found that calix[4]arenes could form intramolecular compounds with solvents retaining their stoichiometric composition during sublimation. Molecules of organic solvents intercalated into calix[4]arene cavities then stabilize the crystal lattice by increasing the enthalpy of sublimation.

On the other hand, some attempts were made to study the adsorption of organic vapors on calixarene thin films prepared by various methods including Langmuir–Blodgett deposition, centrifugation, and self-assembly. It should be noted that the adsorption on films is very often a nonselective process, because the compounds under investigation behave similarly, which is indicative of weak nonspecific interactions between guest molecules and calixarene films [1]. As distinct from crystalline samples, the binding of neutral guests by calixarene films is not a cooperative process accompanied by the formation of a new host–guest phase.

Calixarenes are often used in modeling membrane transport in biological systems because of their unique conformational properties. The size, shape, and electrostatic profile of biological pores are considered basic determining factors of their functioning. The intuitively simple idea of molecular transport through some matrix suggests the presence of channels of the appropriate size limited by van der Waals surfaces. The molecular hydrodynamics, for example, can be considered only when the narrowest part of the channel is at least sufficiently wide for the passage of a separate water molecule. The problems of transport of small mobile molecular particles through a hydrophobic matrix, in particular, water through a thiacalixarene crystal, are discussed in [15, 16]. The conclusion was made that the traditional concept of a porous crystal structure could be incorrect. The formation of clathrates of thiacalixarenes differs essentially from water sorption on activated carbon, where a "step" in the sorption isotherm is supposedly caused by the formation of water clusters [17], and step sorption on zeolites, which are characterized by structure deformation after a significant primary sorption in micropores.

CRYSTAL STRUCTURES OF CALIXARENE COMPLEXES

Many of the calixarenes form crystalline complexes in the solid state with a variety of small molecules, this property having been observed even before the basic structures of the calixarenes were established. For example, *p*-tert-butylcalix[4]arene forms complexes with chloroform, toluene, pyridine [18], benzene, xylene and anisole [19]; *p*-tert-butylcalix[5]arene forms complexes with isopropyl alcohol [20] and acetone [21]; *p*-tert-butylcalix[6]arene forms a complex containing chloroform and methanol [18]; *p*-tert-butylcalix[7]arene forms a complex containing methanol [22]; *p*-tert-butylcalix[8]arene forms a complex with chloroform [18]; *p*-tert-butyldihomooxacalix[4]arene forms a complex with methylene chloride [18]. That the guest molecule is located within the calix of the host molecule is indicated by the X-ray crystallographic pictures of the cyclic tetramer [23] and cyclic pentamer complexes [21]. Derivatives of the calixarenes also frequently show a marked tendency to form complexes. For example, the tetraacetate of *p*-tert-butylcalix[8]arene retains methanol, chloroform, and ethyl acetate far more tightly than does the parent compound [24].

It is well known that calix[4]arenes, particularly when in their *cone* conformation, possess an intramolecular cavity which can host neutral guest molecules of complementary size [25]. Larger calix[n]arenes are also known to form inclusion complexes but because of the usually greater conformational flexibility of these macrocycles and the relatively limited range of studies so far made of them, the factors controlling their inclusion selectivity are not well understood. Calix[4]arenes showed good inclusion properties, in particular with aromatic guests, in the solid state but not in solution [25, 26]. Only recently was it verified that in these media the calixarene cavity can host neutral organic molecules of complementary size. It was also established that the efficiency of the recognition process is strongly determined by the rigidity of the hosts and by the nature of the guest, which should bear acid CH-groups. On this basis it has been hypothesised that specific CH-π interactions stabilise the complexes formed [27-29].

Numerous crystallographic studies have been made of complexes of neutral organic molecules having acidic CH-groups with calix[4]arenes, calix[4]resorcinols and thiacalix[4]arenes in the *cone* conformation. Data retrieved from the Cambridge Structural Database (CSD) concern inclusion resulting fortuitously from the choice of recrystallisation solvent as well as studies deliberately focussed on inclusion complex formation [30]. The influences of molecular symmetry and guest acidity were important issues in the latter group. Solid state structural determinations are frequently complicated by "disorder" problems which have often not been fully resolved [31].

There is the plethora of X-ray crystal structures reported for calixarene complexes, for example with acetonitrile [32, 33-49] and dichloromethane [33, 50-62]. The crystal structures of *p*-tert-butylcalix[4]arene complexes with long-chain guests such as *n*-hexane, 1-pentanol, 1-octanol, dodecan are also well-known [31]. The crystal structures of calix[4]arene *bis*(crown ether)s and their solvent complexes reported up to now are summarized in review article [63].

Chapter 2

CONCURRENT INSERTION OF SOLVENT MOLECULES IN SOLUTIONS

p-Tert-butylcalix[4]arene appears to show little or no selectivity in complexing chloroform and toluene, as indicated by the data on aromatic solvent induced shift (ASIS) and by the fact that strong endo-calix complexes are formed with both of these solvent molecules in the solid state [64]. With any exceptions, for example, complexes with amines, attempts to find guest molecules that can form endo-calix complexes with a calix[4]arene in chloroform solution have been unsuccessful. The problem apparently arises from the fact that the putative guest compounds do not form complexes that are strong enough to "out-compete" the chloroform to an extent sufficient to make the guest detectable by spectral means. The compounds tested included anisole, nitrobenzene, p-xylene, bromobenzene, (trichloromethyl)benzene, (trifluoromethyl)benzene, acetone, tert-butylcyclohexanol, p-tert-butylphenol, benzonitrile, acetonitrile, (trimethylaceto)nitrile, and phenylacetylene. The host molecules used were p-allyl-calix[4]arene and its tetra-p-toluenesulfonate, the latter being chosen because it is known to be fixed in the cone conformation and, therefore, to possess an "enforced cavity" [65]. Concentrations up to 10% of the "guest" molecules were used, and changes in ^{1}H NMR spectral shifts and coalescence temperatures for conformational inversion in the case of the parent calixarene were sought. In no case, however, were significant changes observed. p-2-Hydroxyethylcalix[4]arene and its p-toluenesulfonate are soluble in a 3:1 mixture of acetonitrile and water. It was hoped that complex formation might be demonstrated in these cases as the result of the lipophobic forces that presumably account for the formation of complexes of cyclodextrins in Me$_2$SO solution. However, the ^{1}H NMR spectra of CH_3CN/H_2O solutions of p-2-hydroxyethylcalix[4]arene (or its tosylate) in (trichloromethyl)benzene, toluene, chloroform, benzonitrile, or p-nitrophenol showed no significant changes. Similar results were obtained when 3:1 Me_2SO/H_2O was used as the solvent. The failure to observe complex formation may again arise from competition between the solvent components (H_2O, CH_3CN, Me_2SO) and the putative guest molecules for complexation with the calixarenes.

Representative examples demonstrate that in liquid phase, the solvent alters the stability of the complex in a selective way [66, 67]. Many findings indicate that the solvent effect plays an important role in complexation processes involving ionic or neutral species and macrocyclic ligands in general and calixarenes in particular [68-72].

Danil de Namor's group is very active in the field of the thermodynamic studies of calixarenes which deal with the equilibrium of complex formation with alkali ions and the effects of the solvent [73]. They also studied the thermodynamics of the interactions of calixarenes with ions and neutral molecules. During the investigations they change the applied calixarene host, the guest molecule or ion and the solvent. More than 10 years ago [74] Danil de Namor summarized the problems in the thermodynamics of macrocycles. The role of the reaction media was pointed out in the binding properties of the calixarene esters toward the alkali metal ions. In this field, the properties of the lithium complexes have special importance for constructing high capacity accumulators. The thermodynamics of the dissolution of calixarene derivatives substituted by polar groups at the lower rim was in the focus of Ref. [75]. The solutes are esters with N-di(isopropyl)carbamic acid and ethers with 2-hydroxypyridine moeties substituting all the four lower rim hydroxy groups. The applied polar solvents were nitriles, alcohols and N, N-dimethylformamide among others. The free energy, enthalpy and entropy of solution were calculated from solubility and calorimetric data. In some cases, formation of the solvate layer was observed. The presence of water seriously perturbs the nonaqueous system. This work group summarized the advances on thermodynamics of calixarenes until 1998 in a review article [76]. Properties of about hundred compounds were discussed. Of course, this publication also contains their own results. In this work one can find several very useful data, i.e. tables of thermodynamic data of dissolution of p-tert-butylcalix[n]arenes (n = 4, 6, 8) in solvents of different polarity were published. Another table showed a detailed list of equilibrium constants of lower rim calixarene derivatives with alkali ions in solution. The detailed list of stability constants with neutral species in various solvents was also included. Moreover, one can find a list of the thermodynamic parameters of complexation of calixcrown ligands with metal cations in nonaqueous solvents: equilbrium constants, free energies, enthalpies and entropies of the processes. The last table of this survey contains the properties of the protonation–deprotonation equilibria of upper rim calixarene derivatives in aqueous solutions. During the first years of the present decade several articles were published by this group, dealing with the synthesis of calixarene derivatives, their complex building process and the structure of the complexes. Lower rim calix[4]arene keto derivatives were studied in N,N-dimethylformamide and acetonitrile [77]. Stability constants and standard thermodynamic functions of the complexation with Na^+ were measured. The new 5,11,17,23-tetra-tert-butyl-25,26,27,28-tetra(benzoyl)methoxycalix [4] arene compound was synthetized and its complex with Na^+ and acetonitrile was studied [78]. This molecular complex was isolated and its molecular structure was determined by X-ray diffraction. Thermodynamics of the aforementioned system was reported. Similarly, the compound 5,11,17,23-tetra-tert-butyl[25,27-bis(ethylethanoate)oxy-26,28-bis(ethylthioethoxy)]-calix[4]arene was synthetized [79]. Its hosting capacity was determined toward the metal cations Li^+, Na^+, Ag^+, Ca^{2+}, Cu^{2+}, Hg^{2+}, and Pb^{2+}. The capacity was greater in acetonitrile than in N,N-dimethylformamide and ethanol. The effect of the pendant arms of the host molecule was discussed. Thermodynamics of the complexation process was discussed. Recently, similar measurements were carried out for the investigation of complexation of 4-hydroxypyridine ethers of calix[4]arenes with several metal ions in solutions [80]. The article lists the eqilibrium constants and the changes of the thermodynamic functions (free energy, enthalpy, entropy) during the solvation. Liu et al. [81] published some articles about the thermodynamics of calixarene complexation during the last years. As model host molecules they used water soluble sulfonate derivatives. They studied the complexation process of 5,11,17,23-tetrasulfonato-25,26,27,28-tetrakis(hydroxycarbonylmethoxy)calix[4]arene and 5,11,17,23-tetrasulfonato-thiacalix[4]arene with La^{3+}, Gd^{3+} and Tb^{3+} cations [81]. The host:guest ratio was stoichiometric 1:1 in aqueous solution. The complexations were entropy driven. The large positive entropy change and the smaller positive enthalpy change contribute to the stability of the complexes. In another work [82] they continued the elucidation of the complexation capability of calixarene sulfonates with calix[4]arene tetrasulfonate and thiacalix[4]arene tetrasulfonate. The guest molecules were diazacycloalkanes, i.e. piperazine, homopiperazine and 1,5-diazacyclooctane. The complexes were found enthalpy stabilized. The stabily constants and the changes in the thermodynamic functions were presented. This work was continued using the same host molecules but their guests were in this case pyridinium ions [83]. The highest complex building ability showed the 2,6-dimethylpyridine/2,6-pyridinedicarboxylic acid pair. Similarly, the stability constants and the changes of the thermodynamic function for the complexation processes were determined.

Chapter 3

GAS-PHASE INTERACTIONS OF CALIXARENES WITH MOLECULAR GUESTS

The extensive development of this important class of molecular receptors in the last two decades has expanded considerably the number and variety of substituents carried by the original calixarene structure, especially for the tetrameric calix[4]arene [25, 26, 84-88]. Both the phenolic hydroxyl and the para-substituent were derivatized in many ways, in order to increase the selectivity of the receptor or to modify its flexibility and solubility properties [25, 89]. Despite the enormous work developed throughout the world to synthesize and characterize all these molecular receptors and to investigate their chemistry in solution, very few studies have been undertaken to date to investigate the interaction between them and organic guest molecules and ions in the gas-phase by means of mass spectrometric methods. This lack of interest for the gas-phase chemistry of calixarene- and resorcinarene-cavitands is difficult to understand: even if the practical applications of host-guest chemistry are likely to be developed for the condensed phase, gas-phase studies nevertheless provide a different and interesting perspective for host-guest interactions. As the solvent is absent, no solvation effects can modify the electronic and thermodynamic properties nor the geometrical constraints of supramolecular binding, so that the pure intrinsic interaction between the two counterparts is uncovered, free from any third-body influence.

Much more extensive literature deals with the gas-phase interaction of alkali metal ions with crown ethers, cryptands and other macrocyclic hosts [8, 90]. Another application of mass spectrometry to supramolecular chemistry, quite developed in the past, is the recognition of optically active guests by chiral crown ethers and cyclodextrins [91]. Other recent and rapidly growing fields where mass spectrometry is utilized to characterize complex non-covalent aggregates are, respectively those of organometallic supramolecular architectures [92] and biological protein-substrate and DNA pairing aggregates [93]. Several factors account for the scarcity of work on the gas-phase reactions of calixarenes and resorcinarenes. One reason is that just the basic structures, not the most interesting derivatives, are readily available by commercial sources. Another reason is that complex host structures can not be maintained with ease in an unexcited state in the vapor phase before they react with a charged or neutral guest. The third and more crucial factor is that the interactions between these hosts and neutral molecular guests, which represent the obvious target, are generally very weak and difficult to occur in the gas-phase. It is generally much simpler to pre-form the complexes in solution and then to ionize and isolate them in the gas-phase, as is performed in electrospray ionization (ESI).

The most simple calixarenes are among the molecular receptors that provide too weak interactions to be observed with ease in the gas-phase. The large flexibility of the calixarene structure is emphasized in the gas-phase, where no solvent molecule limits the free conformational changes of the molecule. Consequently, the constraints associated with the formation of host-guest complexes correspond to a strongly negative entropic contribution, resulting in the weakening of the supramolecular interaction. Then very little internal energy is sufficient to dissociate the host-guest complex inside the mass spectrometer, preventing its detection. Even in a specific case studied by Wong and coworkers, where the host had considerable steric hindrance to the free conformational change and the guest carried a positive charge, tert-butylcalix[4]arene proved to be a less effective ligand than crown ethers toward benzylammonium ions [94]. The formation of host-guest complexes in the gas-phase is favored by any form of derivatization, that reduces the flexibility of the calixarene backbone. This decreases the entropic loss associated to the formation of the host-guest complex, making it energetically feasible. A secondary effect of the reduced flexibility of the ligand is its increased selectivity, as the rigid three-dimensional arrangement of its binding sites should complement those of the guest to produce strong interaction. The stiffening of the calixarene structure has been achieved in several ways. One way is to introduce bulky substituents, especially in the lower rim of the molecule, in order to block its structure in the cone conformation. In such a case, it may happen that the calixarene oxygen become quite inaccessible to the candidate guests, restricting the binding properties of the ligand to the π-electrons of its aromatic rings. Thus, other substituents with target binding properties are frequently introduced at the upper rim of the calixarene structure. This solution, was successfully demonstrated by Schalley and coworkers, who studied a series of calix[4]arenes functionalized with urea substituents at their upper rim (i.e., the calixarene opening not containing the hydroxyl oxygen) [92, 95]. Urea substituents are particularly interesting, since they can act both as donor and acceptor in hydrogen bonding. The consequence is that, under ESI conditions, these calixarenes tend to dimerize, forming a two-valve shell held together by hydrogen bonding, in which guest species of molecular dimension can be trapped. Even if the trapped species described in the original work were typically charged (i.e., tetraalkylammonium ions) [92, 95], yielding extremely stable 2:1 complexes, it is not unlikely that also neutral organic molecules capable of hydrogen bonding could be captured inside these supramolecular architectures. More complex ligands, where two calixarene units are covalently linked by means of an hexyl chain produced 1:1 complexes with tetraalkylammonium ions by self-closure of the ligand halves around the guest ion under ESI conditions [95, 96]. When these and other multiple-calixarene ligands were mixed with monomeric urea-substituted calixarenes and alkylammonium salts, the ESI mass spectra showed evidence of stable supramolecular aggregates, containing the three species in specific stoichiometries. These depended on the structure of the coordinating ligand, forming aggregates with up to seven non-covalently bound subunits in a 1:3:3 stoichiometry [95]. Also melamine substituents were introduced into the calixarene structure to form hydrogen-bonded supramolecular assemblies with diethylbarbituric acid and similar monomers exhibiting concurrent donor and acceptor properties [97]. By covalently bridging three calixarene moieties by means of their melamine substituents, large molecular boxes were obtained [98], capable of forming complex molecular aggregates, that were characterized by MALDI-TOF.

Unless time-resolved experiments are performed after ionization, plain ESI mass spectra represent a chemical system where charged host-guest complexes are either pre-formed in solution or are generated at the high-voltage conditions used in ESI. In both cases, the observation of molecular aggregates in the mass spectrum demonstrates the stability of such complexes in the gas-phase, at least for the time-frame of mass analysis. Due to the rather energetic conditions needed inside the mass spectrometer to isolate the analyte ions, the positive finding of a mass peak corresponding to the host-guest complex in the ESI spectrum is generally accepted as a good demonstration of its existence also in solution. In other words, ESI-MS is supposed to provide a reasonable guess of what chemical reactions may take place in solution. Analogous description of mainly the condensed phase chemistry is provided by LSIMS as well as any other technique in which the two counterparts are dissolved together in a liquid matrix and introduced simultaneously inside the ion source of the mass spectrometer. The investigation of gas-phase chemistry requires that the reagents are introduced separately into the mass spectrometer, frequently at different steps of the experiment and using different methods to vaporize them. The most appropriate instruments to perform these studies are time-resolved mass analyzer, capable of trapping ions for long time period, such as Fourier transform ion cyclotron resonance (FTICR) or quadrupole ion-trap mass spectrometers. In these instruments, a substrate ion (for example, the charged host) is first isolated and then allowed to interact with vapors of a neutral reagent species, which is generally pulsed into the reaction chamber. The subsequent mass analysis at various time intervals allows one to determine both the products and the kinetics for their formation.

From the study of the specific gas-phase reactivity of calixarenes with various esters and alcohols, several conclusions of general validity were drawn by authors of [99]. (a) The presence of one or more bridges at the upper rim of the cavity dramatically increases the calixarene ability to form inclusion complexes, provided that these bridges are rather rigid, i.e., all the atoms forming the bridge are linked to at least one unsaturated carbon. (b) The size of the bridge and the nature of its substituents drives the selectivity of the calixarenes toward different guests [100]. (c) Some extreme forms of selectivity that have been experimentally observed may be peculiar to the mass spectrometric context, where the charge location plays the major role.

The possibility of utilizing conventional chemical-ionization (CI) and electron-impact (EI) mass spectrometry as a method for probing gas-phase guest-host chemistry between p-tert-butylcalix[n]arenes (n=4,6) and fluoroaromatics, benzene or 2,2,2-trifluoroethanol as guest molecules is explored in reference [101]. Only with benzene a tiny cluster ion at m/z 727 (p-tert-butylcalix[4]arene+H+PhH)$^+$ and m/z 726 (p-tert-butylcalix[4]arene+PhH)$^+$ observed, which did not increase in intensity when the source temperature was lowered to 120 ^{0} C.

Previous desorption chemical ionization (DCI) complexation studies [102] pointed to a mechanism in which gaseous neutral cavitand molecules encapsulate neutral guests prior to ionization. Mechanisms in which either the charged guest or the host form the guest-host complex could be excluded. Therefore, the lack of complexation in the CI and EI modes may be due to a faster ionization rate of the thermally vaporized neutral species (having more internal energy) relative to their complexation rate. Once the host is ionized it does not complex. The absence of a guest-host cluster in the gas phase from a pre-formed guest-host cluster in the condensed phase shows that the complex undergoes thermal decomplexation prior to (or upon) volatilization/ionization, in line with DCI studies [102]. The mechanistic picture that emerges for gas-phase guest-host complexation between oligomeric macrocycles and neutral guests is that surface-desorbed intact neutral species of low internal energy must be first generated, these subsequently complexing in the DCI plasma [103]. The generated complex is then immediately ionized and detected. EI or CI conditions generate neutral species of higher energy which do not complex.

The inclusion of small neutral organic guests (C_6H_{14}, CH_2Cl_2, CH_3OH) by calix[4]arene receptors was found by ^1H NMR spectroscopy and microanalysis in [104]. The studied calix[4]arenes can form stable intramolecular complexes with solvent molecules which keep the stoichiometric composition without changing under conditions of the sublimation experiment. The saturated vapor pressures of calix[4]arenes and complexes of calix[4]arenes with solvent molecules were determinated for the first time by the Knudsen's effusion method in the wide temperature range. The changing of standard thermodynamic parameters of complexation by transfer process from condensed state to vapor phase was estimated. It was shown that the large flexibility of the calixarene ligand structure corresponds to a strongly negative entropic contribution as well as negative enthalpy term to the Gibbs energy of formation of host-guest complexes in the gas phase.

In this study, the temperature dependences of the saturated vapor pressures of calix[4]arenes and complexes of calix[4]arenes with solvent molecules were determined with the aim to estimate the changing of relative standard thermodynamic parameters of complexation by transfer process from condensed state to vapor phase. 25,27-Dimethoxy-26,28-dihydroxycalix[4] arene (II), 25,27-dimethoxy-calix[4]arene crown-6 (III), 25,27-dimethoxy-5,11,17,23-tetra-p-tert-butylcalix[4]arene crown-5 (IV), 25,27-diethoxy-5,11,17,23-tetra-p-tert-butylcalix[4]arene crown-5 (V), 25,27-bis(benzyloxy)-5,11,17,23-tetra-p-tert-butylcalix[4]arene crown-5 (VI), 25,27-dimethoxy-5,11,17,23-tetra-p-tert-butylcalix[4]arene crown-6 (VII) were synthesized used available from Aldrich 25,26,27,28-tetrahydroxycalix [4]arene (I) (figure 1) according to methods described in ref. [105-108].

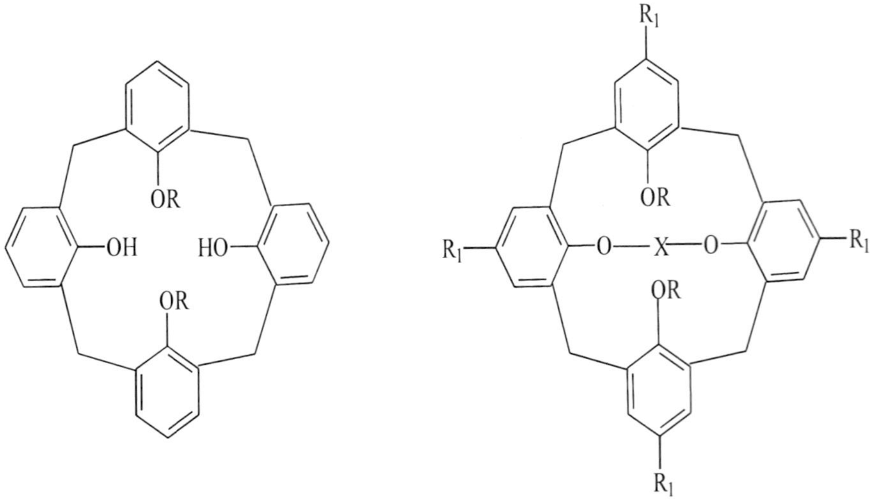

I. R=H
II. R=CH$_3$
III. R=CH$_3$, R$_1$=H, X=CH$_2$CH$_2$(OCH$_2$CH$_2$)$_4$
IV. R=CH$_3$, R$_1$=t-Bu, X=CH$_2$CH$_2$(OCH$_2$CH$_2$)$_3$
V. R=C$_2$H$_5$, R$_1$=t-Bu, X=CH$_2$CH$_2$(OCH$_2$CH$_2$)$_3$
VI. R=CH$_2$C$_6$H$_5$, R$_1$=t-Bu, X=CH$_2$CH$_2$(OCH$_2$CH$_2$)$_3$
VII. R=CH$_3$, R$_1$=t-Bu, X=CH$_2$CH$_2$(OCH$_2$CH$_2$)$_4$

Figure 1. The structural formulas of the calix[4]arenes under investigation [104].

Calix[4]arenes (**I-IV, VII**) are present as a cone structure, compounds (**V, VI**) are in the partial-cone conformation. The complexes (**I**)·C$_6$H$_{14}$·CH$_2$Cl$_2$, (**II**)·C$_6$H$_{14}$, (**III**)·C$_6$H$_{14}$ and (**VII**)·CH$_3$OH were obtained by crystallization from mixture CH$_3$OH/C$_6$H$_{14}$/CH$_2$Cl$_2$. The complex (**I**)·CH$_2$Cl$_2$ was obtained by solution of (**I**) in CH$_2$Cl$_2$ and then by evaporation of the solvent. ^{1}H NMR spectra were recorded with Bruker VC-300 (300 MHz) spectrometer in CDCl$_3$.

Compound (**I**)·CH$_2$Cl$_2$. Found: C, 68.32; H, 5.07; Cl, 13.94. Anal. Calcd for C$_{28}$H$_{24}$O$_4$·CH$_2$Cl$_2$: C, 68.38; H, 5.14; Cl, 13.97. ^{1}H NMR δ 7.97 (s, 4H, OH), 7.39 (d, J = 7.4 Hz, 8H, ArH $_{meta}$), 6.73 (t, J = 7.6 Hz, 4H, ArH $_{para}$), 5.65 (s, 2H, CH$_2$Cl$_2$), 4.33 (s, 4H, ArCH$_2$Ar), 3.42 (s, 4H, ArCH$_2$Ar).

Compound (**I**)·C$_6$H$_{14}$·CH$_2$Cl$_2$. Found: C, 70.59; H, 6.72; Cl, 11.93. Anal. Calcd for C$_{28}$H$_{24}$O$_4$· C$_6$H$_{14}$· CH$_2$Cl$_2$: C, 70.56; H, 6.70; Cl, 11.91. ^{1}H NMR δ 7.95 (s, 4H, OH), 7.41 (d, J = 7.6 Hz, 8H, ArH $_{meta}$), 6.71 (t, J = 7.1 Hz, 4H, ArH $_{para}$), 5.68 (s, 2H, CH$_2$Cl$_2$), 4.31 (s, 4H, ArCH$_2$Ar), 3.40 (s, 4H, ArCH$_2$Ar), 0.98 (m, 8H, C$_6$H$_{14}$), 0.58 (t, 6H, C$_6$H$_{14}$).

Compound **(II)·C$_6$H$_{14}$**. Found: C, 80.30; H, 7.80. Anal. Calcd for C$_{30}$H$_{28}$O$_4$ · C$_6$H$_{14}$: C, 80.27; H, 7.78. ^{1}H NMR δ: 7.71 (s, 2H, OH), 7.82 (d, J = 7.5 Hz, 4H, ArH $_{meta}$), 6.87 (d, J = 7.6 Hz, 4H, ArH $_{meta}$), 6.74 - 6.66 (m, J = 7.2 Hz, 4H, ArH $_{para}$), 4.31 (d, J = 13.1 Hz, 4H, ArCH$_2$Ar), 3.82 (s, 6H, OCH$_3$), 3.41 (d, J = 13.2 Hz, 4H, ArCH$_2$Ar), 0.97 (m, 8H, C$_6$H$_{14}$), 0.55 (t, 6H, C$_6$H$_{14}$).

Compound **(III)·C$_6$H$_{14}$**. Found: C, 74.59; H, 8.11. Anal. Calcd for C$_{40}$H$_{46}$O$_8$ ·C$_6$H$_{14}$: C, 74.56; H, 8.08. ^{1}H HMR δ: 7.80 (d, J = 7.6 Hz, 4H, ArH $_{meta}$), 7.41 (d, J = 7.6 Hz, 4H, ArH $_{meta}$), 6.71 – 6.63 (m, J = 7.1 Hz, 4H, ArH $_{para}$), 4.41 (d, J = 13.2 Hz, 4H, ArCH$_2$Ar), [4.10, 4.02, 3.92, 3.85 (m, 16H, -OCH$_2$CH$_2$O-)], 3.71 (s, 4H, -OCH$_2$CH$_2$O-), 3.12 (s, 6H, OCH$_3$), 3.38 (d, J = 13.2 Hz, 4H, ArCH$_2$Ar), 0.95 (m, 8H, C$_6$H$_{14}$), 0.52 (t, 6H, C$_6$H$_{14}$).

Compound **(VII)·CH$_3$OH**. Found: C, 75.16; H, 9.01. Anal. Calcd for C$_{56}$H$_{78}$O$_8$ ·CH$_3$OH : C, 75.13; H, 8.97. ^{1}H HMR δ: 7.78 (s, 4H, ArH $_{meta}$), 7.39 (s , 4H, ArH $_{meta}$), 4.44 (d, J = 13.1Hz, 4H, ArCH$_2$Ar), [4.15, 4.06, 3.93, 3.81 (m, 16H, -OCH$_2$CH$_2$O-)], 3.64 (s, 4H, -OCH$_2$CH$_2$O-), 3.25(s, 3H, CH$_3$OH), 3.10 (s, 6H, OCH$_3$), 3.40 (d, J = 13.2 Hz, 4H, ArCH$_2$Ar), 1.30 (s, 18H, -C(CH$_3$)$_3$), 1.26 (s, 18H, -C(CH$_3$)$_3$), 1.10 (s, 1H, CH$_3$OH).

Sublimation enthalpy is an important property of the condensed phase as far as this quantity is a macroscopic measure of the magnitude of intermolecular interactions. A variety of experimental techniques have been developed to measure the sublimation enthalpies of solids. In the paper [104] the sublimation experiments were carried out by the Knudsen's effusion method. A weighed in a glass container (± 0.05 mg) sample is placed into the effusion cell. The experimental cell was made of stainless steel with internal volume of about 4 cm^3. The internal diameter of the glass container is about 10 mm, the ratio of sample surface area to effusion orifice area is about from 80 to 300. The design of the experimental cell provides a device for vapor-proof effusion orifice closing during establishing the steady regime of measurements. The temperature of the effusion cell is maintained by means of thermocouples battery with accuracy ± 0.1 ^{0}C. The vapour pressure in the effusion cell is determined by Knudsen's equation (1):

$$P = (\Delta m/\alpha \cdot \beta \cdot S_{eff} \cdot \tau) \cdot (2\pi \cdot R \cdot T/M)^{1/2} \qquad (1)$$

where Δm is a weight loss through an orifice of area S_{eff}, β is the Klausing's factor which takes into account the finite length of the orifice, α is a condensation coefficient, τ is the effusion time, M is the molecular weight, T is a temperature and R is the ideal gas constant.

The effusion equipment was calibrated using naphthalene (m.p.353.43 K) and benzoic acid (m.p.395.5 K). The obtained values of sublimation enthalpies of naphthalene 72.4 ± 0.6 kJ mol^{-1} and benzoic acid 89.8 ± 0.7 kJ mol^{-1} are in good agreement with recommended values [109]. In addition calix[4]arenes (**I-VII**) and their complexes with solvent molecules were purified by sublimation in a high vacuum of 10^{-5} Torr using a temperature gradient furnace. The absence of decomposition and impurities was controlled by ^{1}H NMR spectra. To estimate a condensation coefficient α the sublimation measurements were carried out using two orifices of different effective areas $\beta \cdot S_{eff}$ (2.21 · 10^{-7} m^2 and 8.48 · 10^{-7} m^2) as well as Langmuir method (sublimation from surface of the effusion material). It was found that the measured vapour pressure was independent of both the orifice area and the measurement method (Knudsen or Langmuir). Thus for all compounds under investigation α was equal 1.

The experimentally determined vapor pressure data were described in coordinates lnP versus 1/T by equation (2):

$$\ln P = A + B/T \tag{2}$$

The value of the sublimation enthalpy is calculated by the Clausius-Clapeyron equation (3):

$$\Delta H_{sub}^T = -R \cdot \partial(\ln P)/\partial(1/T) \tag{3}$$

The value of sublimation entropy is calculated as

$$\Delta S_{sub}^T = -\partial(\Delta G_{sub}^T)/\partial T \tag{4}$$

where $\Delta G_{sub}^T = -RT \cdot \ln(P/P_0)$ and $P_0 = 1.013 \cdot 10^5$ Pa. $\quad$ (5)

The results of vapor pressure measurements are plotted in figure 2. The least squares constants A and B corresponding to (2) as well as enthalpies and entropies of sublimation are presented in Table 1.

Table 1. The least squares constants A and B corresponding to equation (2), sublimation enthalpies and entropies of studied compounds

Compound	A	B/1000	ΔH_{sub} T, kJ mol^{-1}	ΔS_{sub} T, J mol^{-1}K^{-1}
(I)	36.5±0.6	-20.1±0.3	167±2	207±5
(I)·CH2Cl2	15±1	-11.7±0.4	98±3	35±5
(I)·C6H14·CH2Cl2	36.6±0.9	-15.7±0.3	131±3	208±8
(II)	17.7±0.6	-9.0±0.2	75±2	71±5
(II)·C6H14	28±1	-14.7±0.5	122±4	134±10
(III)	24.6±0.7	-9.9±0.2	82±2	109±6
(III)·C6H14	25±1	-11.7±0.4	97±3	118±12
(IV)	28.0±0.6	-10.8±0.2	90±2	137±5
(V)	19.2±0.5	-9.5±0.2	79±2	65±4
(VI)	20.5±0.8	-9.3±0.3	78±3	74±7
(VII)	20.8±0.5	-9.3±0.2	78±1	76±3
(VII)·CH3OH	24±1.4	-12.1±0.5	100±4	100±12

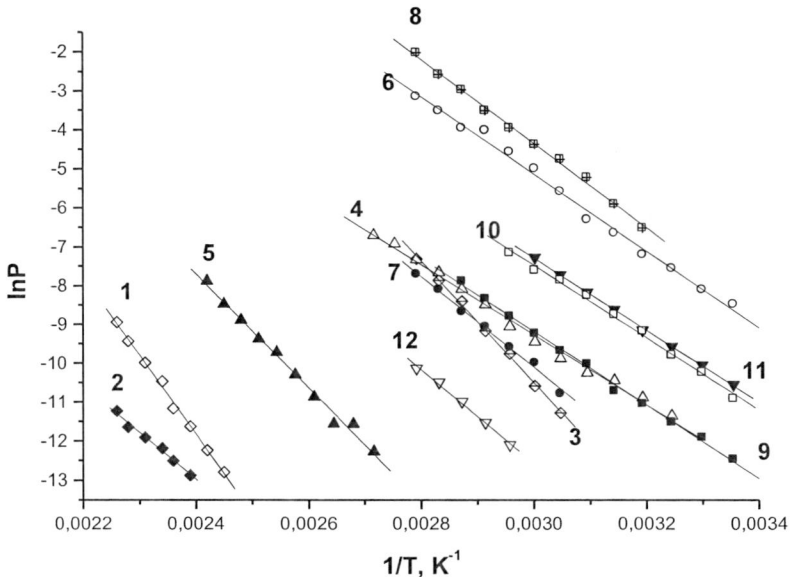

Figure 2. The temperature dependence of the vapor pressures of the studied compounds: 1 - **(I)**; 2 - **(I)**·CH$_2$Cl$_2$; 3 - **(I)**·CH$_2$Cl$_2$·C$_6$H$_{14}$; 4 - **(II)**; 5 - **(II)**·C$_6$H$_{14}$; 6 - **(III)**; 7 - **(III)**·C$_6$H$_{14}$; 8 - **(IV)**; 9 – **(V)**; 10 – **(VI)**; 11 - **(VII)**; 12 - **(VII)**·CH$_3$OH.

It has been shown earlier [13] that the studied compounds can be classified into roughly two groups. The compounds (**I**), (**II**), (**III**) and (**VII**) can form stable intramolecular aggregates (**I**)·CH_2Cl_2, (**I**)·C_6H_{14}·CH_2Cl_2, (**II**)·C_6H_{14}, (**III**)·C_6H_{14} and (**VII**)·CH_3OH which keep the stoichiometric composition without changing under conditions of the sublimation experiment. The calix[4]arenes (**IV**), (**V**) and (**VI**) do not form stable host-guest complexes with the solvent molecules. The calix[4]arenes (**V**) and (**VI**) are in the partial-cone conformation and their potential cavities are full with their own groups ("endo-positioned" ethoxy and benzyloxy groups, respectively) [105]. The structure difference of the calix[4]arenes (**IV**) and (**VII**) is the additional group OCH_2CH_2 in the crown polyether ring of (**VII**). Calix[4]arene (**VII**) forms the stable intramolecular aggregate with the solvent molecule as a result of the increase of the length of the polyether ring of (**VII**). The inclusions of organic molecules to the cavities of calixarenes (**II**), (**III**) and (**VII**) stabilize the crystal lattice energies increasing the values of the sublimation enthalpies. The sublimation enthalpy of ligand (**I**) is higher than enthalpies of (**I**)·CH_2Cl_2 and (**I**)·C_6H_{14}·CH_2Cl_2. The complex (**I**)·C_6H_{14} is unstable and decomposed under sublimation. The large flexibility of the calixarene structure is emphasized in the gas phase, where no solvent molecule limits the free conformational changes of the molecule. Consequently, the constraints associated with the formation of host-guest complexes correspond to a strongly negative entropic contribution, resulting in the weakening of the supramolecular interaction. The formation of host-guest complexes in the gas phase is favored by any form of derivatization that reduces the flexibility of the calixarene backbone. This decreases the entropic loss associated to the formation of the host-guest complex, making it energetically feasible. A secondary effect of the reduced flexibility of the ligand is its increased selectivity, as the rigid three-dimensional arrangement of its binding sites should complement those of the guest to produce strong interaction. The stiffening of the calixarene structure has been achieved in several ways. One way is to introduce bulky substituents, especially in the lower rim of the molecule, in order to block its structure in the cone conformation. In such a case, it may happen that the calixarene oxygen become quite inaccessible to the candidate guests, restricting the binding properties of the ligand to the π-electrons of its aromatic rings.

The unique property of complexes of calixarenes with solvent molecules to keep the stoichiometric composition without changing under conditions of the sublimation experiment efforts an opportunity to estimate the thermodynamic functions of complexation using data on temperature dependence of saturated vapour pressures. The thermodynamic cycle for the transfer process of calixarene, solvent and the complex of calixarene with solvent from the condensed state to the gas phase is considered to be:

$$nHost_g + mGuest_g \rightarrow nHost \cdot mGuest_g$$

↑ ↑ ↑

$$nHost_s + mGuest_l \rightarrow nHost \cdot mGuest_s$$

The appropriate enthalpies correspond to the stages:

$nHost_s \rightarrow nHost_g + \Delta H^0_1$
$mGuest_l \rightarrow mGuest_g + \Delta H^0_2$
$nHost \cdot mGuest_s \rightarrow nHost \cdot mGuest_g + \Delta H^0_3$
$nHost_s + mGuest_l \rightarrow nHost \cdot mGuest_s + \Delta H^0_{complex1}$
$nHost_g + mGuest_g \rightarrow nHost \cdot mGuest_g + \Delta H^0_{complex2}$

The relative binding enthalpy is computed by the difference:

$$\Delta(\Delta H^0_{complex}) = \Delta H^0_{complex2} - \Delta H^0_{complex1} = \Delta H^0_3 - (\Delta H^0_1 + \Delta H^0_2)$$

Similarly, assuming that the system is reversible:

$$\Delta(\Delta G^0_{complex}) = \Delta G^0_{complex2} - \Delta G^0_{complex1} = \Delta G^0_3 - (\Delta G^0_1 + \Delta G^0_2)$$

where, *Host* is a calixarene; *Guest* is a solvent; *nHost·mGuest* is a complex of the calixarene with the solvent; n and m characterize the stoichiometry of the complex; the subscript signs s, l correspond to the condensed state, g corresponds to the gas phase, respectively; $\Delta H^0_{1,2,3}$ and $\Delta G^0_{1,2,3}$ are the standard enthalpies and Gibbs energies of vapour formation of calixarene, solvent and the complex of the calixarene with the solvent, respectively; $\Delta H^0_{complex1}$, $\Delta H^0_{complex2}$ and $\Delta G^0_{complex1}$, $\Delta G^0_{complex2}$ are the standard enthalpies and Gibbs energies of complexation in the condensed state and in the gas phase, respectively.

Thus the certain values of $\Delta H^0_{1,2,3}$ and $\Delta G^0_{1,2,3}$ allow us to calculate the relative binding enthalpies $\Delta(\Delta H^0_{complex})$ and Gibbs energies $\Delta(\Delta G^0_{complex})$ of complexation by transfer process from the condensed state to the gas phase and the corresponding changing of entropy $T\Delta(\Delta S^0_{complex}) = \Delta(\Delta H^0_{complex}) - \Delta(\Delta G^0_{complex})$. The relative standard thermodynamic functions of complexation

$\Delta(\Delta H^0_{complex})$, $\Delta(\Delta G^0_{complex})$, $T\Delta(\Delta S^0_{complex})$ for the complexes $(I) \cdot CH_2Cl_2$, $(I) \cdot CH_2Cl_2 \cdot C_6H_{14}$, $(II) \cdot C_6H_{14}$, $(III) \cdot C_6H_{14}$ and $(VII) \cdot CH_3OH$ are presented in Table 2.

Table 2. The relative standard thermodynamic functions of complexation by transfer process from the condensed state to the gas phase

Compound	$\Delta(\Delta G^0_{complex})$, kJ mol^{-1}	$\Delta(\Delta H^0_{complex})$, kJ mol^{-1}	$T\Delta(\Delta S^0_{complex})$, kJ mol^{-1}
$(I) \cdot CH_2Cl_2$	-19	-98	-80
$(I) \cdot CH_2Cl_2 \cdot C_6H_{14}$ [a]	-43	-96	-54
$(II) \cdot C_6H_{14}$	19	16	-3
$(III) \cdot C_6H_{14}$	10	-16	-26
$(VII) \cdot CH_3OH$	11	-15	-26
$(I) \cdot CH_2Cl_2 \cdot C_6H_{14}$ [b]	-24	2	26

[a] complexation process: $(I) + CH_2Cl_2 + C_6H_{14} \rightarrow (I) \cdot CH_2Cl_2 \cdot C_6H_{14}$.
[b] complexation process: $(I) \cdot CH_2Cl_2 + C_6H_{14} \rightarrow (I) \cdot CH_2Cl_2 \cdot C_6H_{14}$.

The sublimation enthalpies at the mean temperatures of measurements were assumed as the standard sublimation enthalpies of calixarenes and complexes of calixarenes with solvents (Table 1). The evaporation enthalpies of C_6H_{14}, CH_2Cl_2 and CH_3OH at 298.15 K are 7.4; 6.8 and 8.94 kcal mol^{-1}, respectively. The standard sublimation Gibbs energies ΔG^0 of calixarenes and their complexes were calculated by equation (5) using vapour pressures values obtained by extrapolation of the temperature dependence of vapour pressures to 298.15 K (Table 1). The standard evaporation Gibbs energies ΔG^0 of solvents were calculated using vapour pressures values at 298.15 K.

As the sublimation entropy characterizes the free conformational changes of the molecule by the transfer process to the gas phase, it is possible to estimate the correlation between flexibility of the calixarene structure and the complexation properties.

Grootenhuis et al. [110] assessed some of the structural and energetical properties of calix[4]arenes by various computational methods. They found that the preferred conformation of calix[4]arene depends on the number and the positions of the substituents on the oxygen atoms and is mainly determined by electrostatic interactions. In a number of cases the conformation with the lowest calculated energy is different from the conformation found in solution and in the solid state. In order to study conformational interconversions the authors [110] carried out molecular dynamics simulations on one isolated molecule of 25,26,27,28-tetrahydroxy-5,11,17,23-tetra-p-methylcalix[4]arene at temperatures of 300, 600, and 800 K for 50 ps. Inspection of MD movies suggested that at 300 K the motions of the two phenol moieties opposite to each other were initially strongly coupled when the calixarene was in its *cone* conformation. After 35 ps one of the phenol moieties did break its hydrogen bonds with the other phenols and flipped through the cavity, resulting in the *partial cone* conformation. After adopting the *partial cone* conformation for a few picoseconds the molecule started interconverting between the *1,2-* and *1,3- alternate* conformations and the *partial cone*. At 600 K only one conformational interconversion was observed in the 50-ps trajectory. However, the amplitude of the coupled movements of the phenolic moieties was much larger at this temperature. At 800 K an almost continuous interconversion of the calixarene was observed. It is quite clear from the MD movies that the strongly coupled movement of the two pairs of opposite phenol moieties favors the conformational transitions. So compound (**I**) is conformationally mobile in the gas phase and can interconvert between the *cone, partial cone, 1,2-alternate, 1,3-alternate* conformations, although in the solid state it is present exclusively as a *cone* structure due to strong intramolecular hydrogen bonding [105]. Consequently, the transfer of calixarene (**I**) from the solid state to the gas phase must be accompanied by high value of sublimation entropy.

The conformational analysis of calixarene (**II**) is described in [110]. For the dimethyl ether (**II**) none of the conformations is clearly preferred in the calculations. The *cone* conformation is not anymore the only conformation in which all possible (in this case two) *H* bonds can be accommodated, and therefore, other conformations that have a more favorable VDW or bonded energy or have a lower electrostatic repulsion between the oxygen atoms become important. On the other hand, the dimethyl ether (**II**) can in principle assume every possible conformation due to its flexibility in the gas phase. Thus the sublimation process should be characterized in this case by low value of sublimation entropy.

The flexibility of (**I**)·CH_2Cl_2 (which is considered as the ligand in the complexation process (**I**)·CH_2Cl_2 + C_6H_{14} → (**I**)·CH_2Cl_2·C_6H_{14}) is constrained greatly as a result of the formation of host-guest complex. The presence of the crown polyether ring at the lower rim of the cavity reduces the flexibility of the calixarene structures (**III**) and (**VII**) in comparison with (**I**) and (**II**). *tert*–Butyl groups introduced at the upper rim of the calixarene structure (**VII**) reduce the flexibility of the calixarene backbone as compared to (**III**).

The relative standard thermodynamic functions of complexation by transfer process from the condensed state to the gas phase versus sublimation entropy of calixarene ligands, $T\Delta S^0_{subl}$, are shown in figure 3.

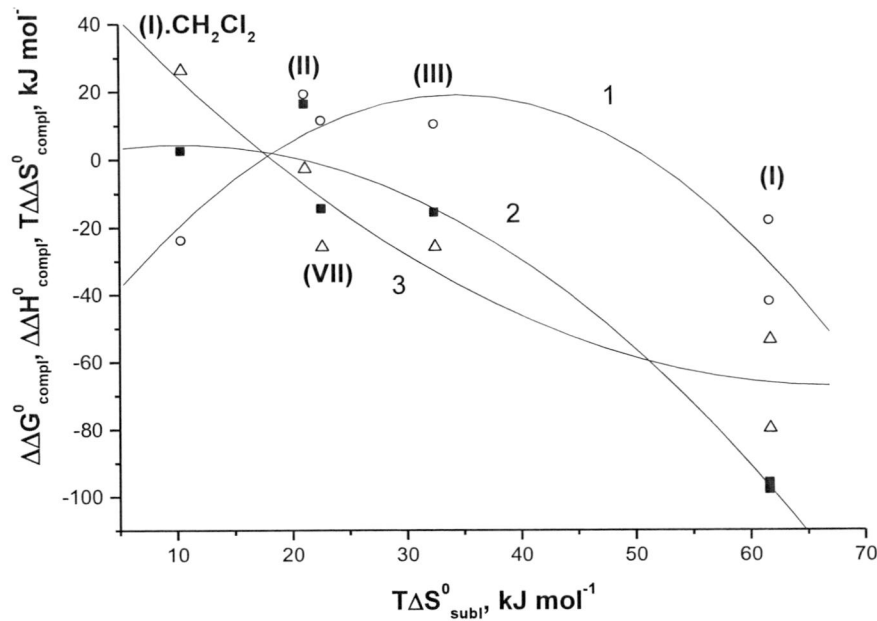

Figure 3. The relative standard thermodynamic functions of complexation by transfer process from the condensed state to the gas phase. $T\Delta S^0_{subl}$ is sublimation entropy of calixarene ligands (**I**), (**II**), (**III**), (**VII**) and (**I**)·CH$_2$Cl$_2$; 1 – $\Delta(\Delta G^0_{compl})$; 2 – $\Delta(\Delta H^0_{compl})$; 3 – $T\Delta(\Delta S^0_{compl})$.

It is obvious that the smaller value of sublimation entropy ($T\Delta S^0_{subl}$ values decrease in order (**I**) > (**III**) > (**VII**) > (**II**) > (**I**)·CH$_2$Cl$_2$) corresponds to the more positive entropic term $T\Delta(\Delta S^0_{complex})$ to the relative standard Gibbs energy of complexation $\Delta(\Delta G^0_{complex})$. On the other hand, the negative enthalpy contribution $\Delta(\Delta H^0_{complex})$ to the Gibbs energy of complex formation $\Delta(\Delta G^0_{complex})$ increases with increasing of $T\Delta S^0_{subl}$. As a result the flexibility dependence of the relative standard Gibbs energy of complexation ($\Delta(\Delta G^0_{complex})$ versus $T\Delta S^0_{subl}$) is an extremal function. The negative values of $\Delta(\Delta G^0_{complex})$ correspond to the maximal and minimal conformational changes of the calixarene ligand structures by transfer process from the solid state to the gas phase. The entropic contribution to the relative free energy of complexation dominates when the flexibility of calixarene is reduced whereas the enthalpy term prevails when the flexibility is large. The complex formation (**I**)·CH$_2$Cl$_2$ + C$_6$H$_{14}$ → (**I**)·CH$_2$Cl$_2$·C$_6$H$_{14}$ is accompanied by negligible enthalpy effect and is governed by the entropic contribution to the relative

Gibbs energy of the complexation process. The values of the relative Gibbs energy of complexation of (**II**), (**III**) and (**VII**) are positive since negative values of $T\Delta(\Delta S^0_{complex})$ are not balanced by negative values of $\Delta(\Delta H^0_{compl})$. Hence the preferred complexation occurs in the condensed state coming out of insufficient rigid structures of calixarene ligands (**II**), (**III**) and (**VII**). Thus it must be taken into consideration that the large flexibility of the calixarene ligand structure corresponds to a strongly negative entropic contribution as well as negative enthalpy term to the Gibbs energy of formation of host-guest complexes in the gas phase.

More than two decades after their introduction, calixarene and resorcinarene cavitands are nowadays investigated for their possible practical applications. To this respect, gas-phase interaction studies have limited direct impact on applications, and should still be regarded chiefly as fundamental studies. However, they provide a unique perspective and information, since the dynamic character of the gas-phase interaction makes the attractive forces established between host and guest depend uniquely on their relative structures and the intrinsic energy of the system.

Moreover, the major source of interference, that is the solvent, is absent in gas-phase studies. The simplicity of the reacting system has the potential of revealing the specific nature and strength of each supramolecular binding. Thus, the thermodynamic parameters associated to subtle structural attributes such as, for example, synergic multiple binding, conformational dynamics and steric constraints can be measured. In order to undertake these studies, it is convenient to establish strong cooperation between synthetic chemists and mass spectroscopists, and to have access to instruments capable of time-resolved experiments, particularly FTICR mass spectrometers. But, above all, it is essential that the scientist investigating supramolecular architectures become aware of the significance of gas-phase studies in elucidating the intrinsic nature of supramolecular interactions.

Chapter 4

ADSORPTION OF ORGANIC VAPORS WITHIN CALIXARENE THIN FILMS

Atmospheric detection of odorant vapors of organic compounds is one of the most important problems of environment monitoring. A number of vapor sensors and sensor arrays have been developed during the last few years. The majority of these sensors are based on the use of different polymer coatings and employ several transduction techniques, such as, using a mass sensitive quartz crystal microbalance (QCM) [111-115] and surface acoustic wave (SAW) [116, 117] devices, chemical field effect transistors (ChemFETs) [118], conductivity [111, 119-121] and interdigitated capacitance [122] measurements.

Several attempts to study the adsorption of organic vapors within thin calixarene films formed with different techniques, including Langmuir-Blodgett (LB) film deposition, spin coating and self-assembly, have been made [123-129]. In publications [128, 129] it has been shown that the vapors of benzene and toluene, as well as some hydrocarbons (hexane), can be adsorbed at calixarene LB films. This adsorption process is very fast, and full recovery of the film has been observed after flushing with clean air. It has to be pointed out, however, that the detected vapors were of a high concentration (a few percent in volume) and the adsorption was not selective since all vapors studied yielded a similar response. These effects are attributed to weak and non-specific interactions between guest molecules and the calixarene LB film. It was also shown that the adsorption of organic vapors occurs in the whole bulk of the LB films, and that the number of adsorbed molecules is much higher than the number of calixarene molecules [128]. The proposed mechanism of adsorption included swelling of the film and even condensation of adsorbate within the film. The swelling of the film has been confirmed directly by ellipsometry and surface plasmon resonance (SPR) measurements [128, 129], but the mechanism of adsorption is still unclear.

Adsorption of vapors of benzene, toluene, p-xylene, aniline, hexane and chloroform in LB films of two calix[4]resorcinarene derivatives was studied in situ using QCM, ellipsometry and SPR techniques [130]. Isotherms of adsorption obtained by both QCM and SPR show that the adsorption ability depends on the condensed vapor pressures of the adsorbates rather than on a structural coincidence between host cavities and guest molecules. The results were interpreted in terms of capillary condensation of organic vapors in the nanoporous matrix of calixarene LB films accompanied by film swelling. Ellipsometric measurements show changes of both the thickness and refractive index of the LB films cased by adsorption, and thus confirm condensation and further accumulation of liquid adsorbate within the film matrix.

In the paper [131] a systematic study is reported of modified calixarenes with various side groups and various sizes to optimize their recognition properties towards specific molecules. Thin films of calix[4]arenes, calix[6]arenes, and calix[8] arenes have been prepared by thermal evaporation onto Au electrodes of quartz oscillators. Since polar O-containing groups are present in para-positions of the aromatic rings, double-layer structures can be formed in thin calix arene films, similar to the geometric structure of toluene/ p-tert-butylcalix[4]arene crystals [84]. The first set of experiments in this study deals with the determination of thermodynamic and kinetic parameters which describe the interaction of thin calixarene layers with organic solvent molecules. Perchloroethylene (C_2Cl_4) was chosen as a solvent with particular relevance in environmental control. The parameters include the equilibrium surface and bulk concentrations, molar heats of absorption, diffusion coefficients, and activation energies of diffusion. They are obtained from thickness, partial pressure, temperature, and time dependent measurements of mass changes which occur during the solvent/calixarene interaction. These parameters are also important from a practical viewpoint in characterizing and comparing the different properties of calixarenes for specific applications as chemical sensors. In a second set of experiments, activation energies were determined for desorption and diffusion of C_2Cl_4 molecules by means of thermodesorption spectroscopy (TDS). The latter is performed under ultra-high-vacuum conditions with well defined calixarene layers prepared in-situ by Knudsen evaporation onto gold substrates. Finally experimental results obtained for different solvents were compared (i.e. perchloroethylene (C_2Cl_4), chloroform ($CHCl_3$), benzene (C_6H_6), and toluene ($C_6H_5CH_3$)) with theoretical force field calculations of geometric structures and binding energies of molecule/calixarene complexes. From thickness dependent measurements of mass changes during subsequent exposure of different solvents to these films surface effects were separated from bulk effects in the solvent/calixarene interactions. Thermodesorption spectra give additional evidence for diffusion processes occuring from surface to subsurface sites even at low temperatures T~ 240 K. Force field calculations indicate strong key-lock interactions between individual solvent molecules with molecular cavities of calix[4]arenes. The larger cavities of calix[6]arenes and calix[8]arenes form supramolecular units which capture two or more small organic solvent molecules.

Chapter 5

MOLECULAR RECOGNITION OF ORGANIC GUEST VAPORS BY SOLID CALIXARENES

Molecular recognition in solid-phase host-guest binding is generally a more complex phenomenon than solvation selectivity in liquids. For liquid solutions the linear structure-energy relationship is usually observed in the absence of specific solute-solvent coordination or for simple solvent-solute interaction with the single donor-acceptor or H-bond [132]. Molecular recognition in the supramolecular systems is based on host-guest structural complementarity [133], which is essentially a nonlinear property. This nonlinearity can be found in the discrete changes of the stoichiometry of host-guest inclusion compounds with variation of the guest molecular structure [134-138]. More guest and host structural features are relevant for the guest inclusion than for solvation of organic compounds. For liquid solutions the knowledge of the solute and solvent molecular group composition is often enough for the prediction of solvation parameters [139]. The guest inclusion parameters depend, besides, on the host shape and symmetry, conformational flexibility/rigidity, and configuration of the H-bonding network [134, 140]. The knowledge of parameters of guest molecular recognition by solid hosts can help to improve the host molecular design, especially when the full structural picture of the guest inclusion is not available. An effective strategy for the study of the host structural influence on guest inclusion can be the examination of the structure-affinity relationships for hosts with a few different structural features. This approach was realized consistently only for the stoichiometry of the inclusion compounds [134, 135, 141-143]. The ability of the inclusion compound to crystallize from the host solution in liquid guest is often used as a qualitative host-guest affinity parameter [134, 135, 141-143].

Another related parameter is the difference between the guest boiling temperature and the decomposition temperature of the inclusion compounds [144]. Estimation of the guest binding energy is not a simple problem for the inclusion process due to its high cooperativity [145-148], sensitivity to temperature [149-151], and the presence of other components [146, 152-156]. These effects, if not controlled, can considerably distort the relationship between the obtained inclusion parameters and the host and guest molecular structure.

The structure-affinity relationships were studied for the guest inclusion parameters of solid tert-butylthiacalix-[4]arene and tert-butylcalix[4]arene in [157, 158]. The inclusion stoichiometry and inclusion free energy were calculated by the sorption isotherms obtained for guest vapor-solid host systems by the static method of headspace gas chromatographic analysis at 298 K. The obtained sorption isotherms have an inclusion threshold for guest thermodynamic activity corresponding to the phase transition between the initial host phase and the phase of inclusion compound.

The influence of the calixarene macrocycle size on the thermodynamic parameters of inclusion formation in organic guest vapor-solid host systems was studied in the series of tert- butylcalix[4]arene, tert-butylcalix[6]arene, and tert-butylcalix[8]arene [159]. For this purpose, sorption isotherms of a guest vapor by a solid host were determined using the static method of headspace gas chromatographic analysis. Besides, the stoichiometry and decomposition temperatures of saturated clathrates formed in these systems were determined using thermogravimetry. The compositions of some of these clathrates differ substantially from those of clathrates crystallized from a host solution in a liquid guest.

The special features of water interaction with classic calixarenes were studied [160] by measuring water sorption isotherms on seven powdered calix[4]arenes (I–VII) (figure 1) at 298 K by the isopiestic method (storage in an desiccator at controlled humidity). Isopiestic experiments were performed at relative pressures p/p_0 from 0.15 to 1. Calixarene powders (fractions with particle size 0.16–0.28 mm separated by screening through calibrated sieves) with weight about 500 mg were dried in a vacuum at 338–343 K to a constant weight and brought to isopiestic equilibrium with aqueous solutions of sulfuric acid with a given concentration at the known vapor pressure over solution. The experiments were performed in an air thermostat; the temperature was controlled to an accuracy of ± 0.1 K. The time of the establishment of equilibrium was from 7 to 21 days. The content of sorbed water was determined gravimetrically. Figure 4 shows water sorption isotherms on calixarenes at 298 K.

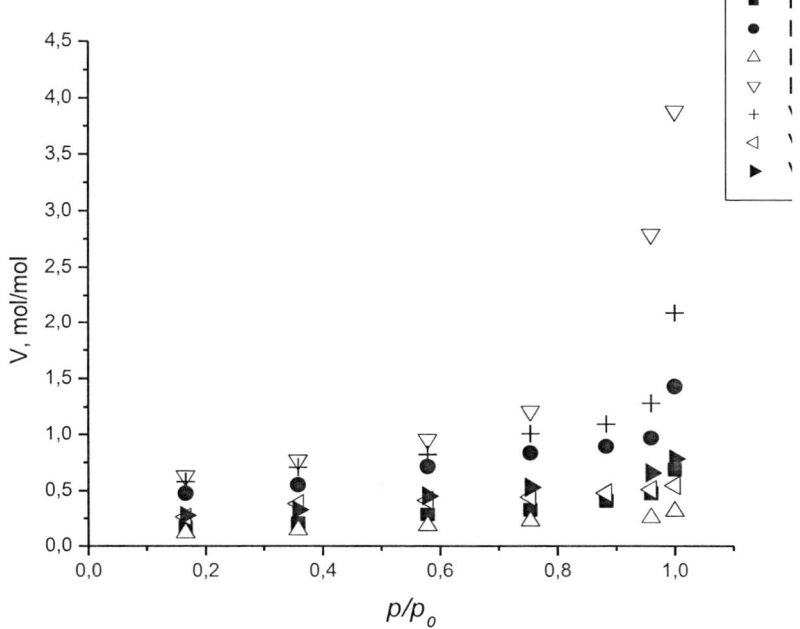

Figure 4. Isotherms of water sorption on calixarenes I–VII at 298 K.

As can be seen from the figure, all isotherms are similar and show a pronounced trend to a sharp increase in water sorption as p/p_0 approximates 1. The complete coincidence between the sorption and desorption branches of isotherms and the absence of hysteresis are indicative of an insignificant porosity of crystalline calixarenes. A comparative analysis of sorption isotherms and the structural formulas of calixarenes shows that the water sorption value is independent of the number of potential sorption sites (oxygen atoms in calixarene molecules). The maximum number of oxygen atoms is characteristic of calixarenes **III** and **VII**, and maximum water sorption is observed for calixarenes **IV** and **V**. Maximum water sorption on various calixarene samples is 0.5–1 mol/mol on average, except samples **V** and **IV** characterized by water sorption of 2 and 4 mol/mol calixarene, respectively. An analysis of the isotherms showed that the experimental data were linearized in the coordinates of the Harkins–Jura equation. Figure 5 demonstrates isotherms of water sorption on the calixarenes in the coordinates of the Harkins–Jura equation.

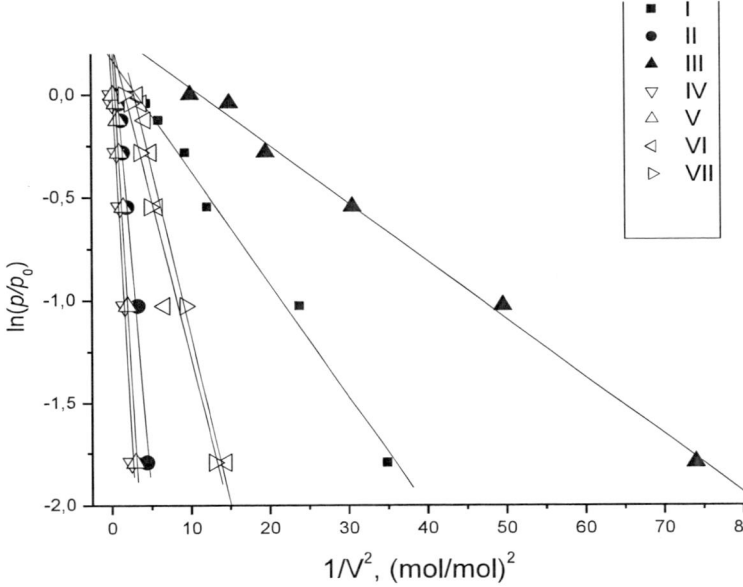

Figure 5. Isotherms of water sorption on calixarenes I–VII in the coordinates of the Harkins–Jura equation.

By analyzing the phase diagrams of surface layers constructed from the experimental isotherms of gas and vapor adsorption on solid adsorbents, Harkins and Jura proved that the state of two-dimensional surface films was similar to that of adsorption layers on solid surface [161]. The works by Harkins and Jura confirmed that there were well-defined regions of the curves of compression of the surface layers of solid adsorbents similar in properties to various film phases, for example, oil on water. Each of five phases of oil-on-water films demonstrates a relation between film pressure π and molecular surface area σ characteristic of the given phase only. Harkins and Jura showed that the isothermal equation of state for a condensed film is described by the equation

$$\pi = b - a\sigma, \qquad (6)$$

where π is the surface pressure of the film, σ is the molecular surface area, and *a* and *b* are constants. This equation can be transformed as

$$\ln(p/p_0) = B - A/V^2, \qquad (7)$$

where p/p_0 is the relative pressure of adsorbate vapors, *A* and *B* are constants, and *V* is the amount of the adsorbed substance.

According to the data by Harkins and Jura, the surface area of a solid can easily be calculated from a vapor adsorption isotherm provided that the adsorbed film forms a condensed phase along part of the isotherm; the surface area of a solid can be calculated by the equation

$$\Sigma = kA^{1/2}, \qquad (8)$$

where Σ is the surface area, *k* is the constant depending on the adsorbate used, and *A* is the constant of Eq. (7). For water at 25°C, $k = 3.83$ [161]. The surface areas (m^2/g) of calixarenes **I–VII** calculated using the Harkins–Jura approach are given below: 2.11 (**I**), 5.813 (**II**), 0.984 (**III**), 3.641 (**IV**), 3.452 (**V**), 1.423 (**VI**), and 1.609 (**VII**).

An analysis of the X-ray diffraction data on single crystals of **IV**, **V**, and **VI** [105] shows that the enhanced sorption on samples **IV** and **V** can be related to the presence in the crystal lattices of a developed network of through channels favoring the penetration of water molecules into hydrophobic crystals (figure 6).

Water is not adsorbed on specific sorption sites but forms a condensed film, which, in the presence of a sufficient number of branched channels, can form a polymolecular film. As distinct from classic calixarenes, water absorption by thiacalixarene follows a complex mechanism including rotations of *tert*-butyl groups accompanied by deformation of the host molecule with the removal of peripheral aromatic rings that form the cavity [16]. Such a deformation most likely involves rotation around C–S thioether bonds. It is well known that, compared with classic calixarenes, thiacalixarenes are characterized by improved mobility in solutions and conformational peculiarities in the solid state [162]. The improved conformational mobility of solid thiacalixarenes is probably responsible for the intercalation of water molecules into a calixarene crystal without the obvious presence of suitable channels [16]. According to [16], the three-dimensional lattice and molecular cavities in the structure of thiacalixarene are the determining factors for water diffusion. Although the cavities do not merge to form channels, water molecules can probably use these cavities as steps in their motion through the lattice. Therefore, the mechanism of water transport should involve jumps of water molecules between cavities until a thermodynamically favorable arrangement is found. In the case of classic calixarenes **I–VII** without enhanced conformational mobility in the solid state, water cannot supposedly be intercalated into hydrophobic cavities, and is adsorbed on the surface of crystals or through pores with fairly large diameters to produce a condensed film.

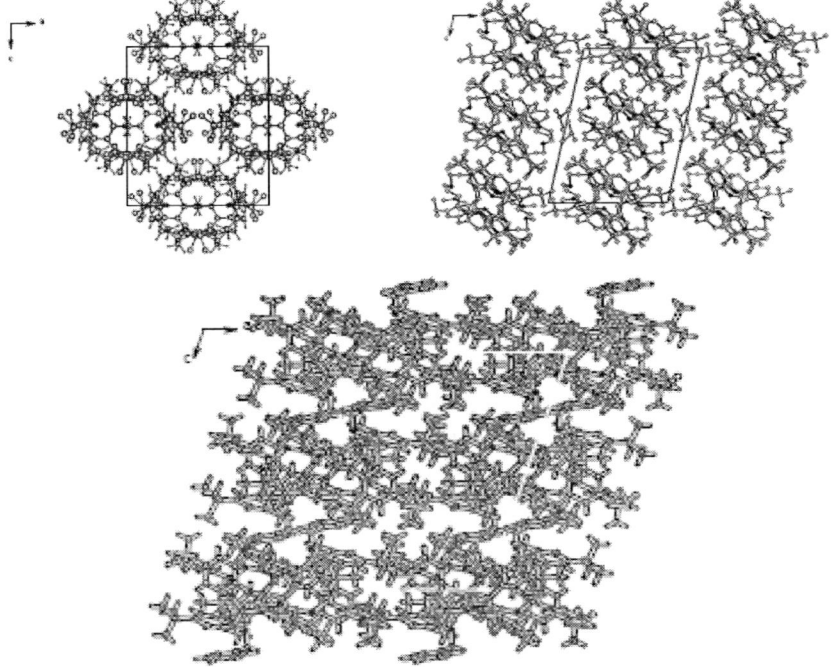

Figure 6. Porous structure of calixarenes IV, V, and VI.

Binding of small neutral guest with solid monodeoxycalix[4]arene host was carried out in references [163, 164]. It was found that the solid apo-host can bind gaseous small organic guest; the guest selectivity in the cavity of solid apo-host is closely related to the free energy of complexation in solution. Structure elucidation of the inclusion complexes of monodeoxycalix[4]arene with small organic guests was carried out by X-ray crystallographic analysis and molecular dynamic simulation. Although the guest moves rapidly in the host cavity, the time averaged structure resembles the one obtained by the X-ray crystallography. Chemical shift simulation succeeded to reproduce the observed complexation induced shift.

It was shown that a potassium salt of *p*-tert-butylthiacalix[6]arene shows highly-extensive coordination nature to give rise to zeolitic crystal which has huge cavity with the widest span of ca. 19 Å and is capable of crystalline phase guest-addition and –removal [165]. This work suggests a potential utility of the bridging sulfur moieties of thiacalixarene for construction of extensively-coordinated structures in the crystalline state. The sulfur-bridged cyclic hexamer drastically differs from a methylene-bridged counterpart *p*-tert-butylcalix[6]arene in that a former can bind as much as four potassium cations at once, while a latter usually forms the complexes with not more than two alkali metal ions [166, 167]. The obtained highly-coordinated crystal may be regarded as intermediate between pure organic crystals involving porosity based on phase transition [148] upon guest binding and metal-coordinated ones possessing permanent or genuine porosity.

The solid state inclusion of various organic solvent molecules in *p*-tert-butylcalix[4]arene and *p*-tert-butylcalix[6]arene has been studied by thermal gravimetry and electron impact mass spectrometry (EI-MS) in [168]. The host–guest ratio varies from 1:2 to 4:1 and the nature of included guest has been determined by EI-MS. Thermal gravimetric analysis of solvent–*p*-tert-butylcalix[*n*]arene complexes gives a qualitative order of intramolecular forces involved. Structural information obtained by cross-polarization magic angle spinning (CP-MAS) 13C NMR spectroscopy is in good agreement with known data from single crystal X-ray diffraction analysis.

In [169] the authors report evidence for an inclusion complex between carbon dioxide and *p*-tert-butylcalix[4]arene. The complex was studied with infrared spectroscopy, single-crystal X-ray diffraction, solid-state NMR, and thermogravimetric analysis. Results indicate that 70% of the *p*-tert-butylcalix[4]arene cavities could be occupied by a CO_2 molecule following exposure a t 30 MPa and 40 °C.

With recent developments in materials chemistry and nanotechnology, extensive studies have been undertaken in the search for stable nano- or microporous networks. Most of these networks utilize coordination polymers, the porosity of which can be programmed depending on the applications. A variety of metal organic frameworks (MOFs) has been designed for gas storage and transport [170-179]. MOFs were found to effectively absorb N_2, O_2, Ar, CO_2, N_2O, H_2 and CH_4. The remarkable ability of certain MOFs in sorption of H_2 and CH_4 makes them very attractive candidates for vehicular gas storage. Much less effort has been devoted to assessment of pure organic solids as gas sorbents since organic molecules typically adhere to close-packing principles and do not afford porous structures. Calixarenes offer a remarkable exception. Atwood and co-workers showed that CH_4, CF_4, C_2F_6, CF_3Br and other low-boiling halogenated alkanes could be reversibly entrapped and retained within the lattice voids of a crystalline calix[4]arene framework [180]. Such gas-storing crystals appeared to be extremely stable and release their guests only at elevated temperatures, several hundreds of ^0C above their boiling points. Ripmeester discovered that the calix[4]arene cavities in such crystals are directly involved in the gas complexation [181]. The Atwood team further demonstrated that p-tert-butylcalix[4]arene dimerizes in a crystalline phase into a hourglass-shaped cavity, capable of gas entrapment [182]. These crystals soak up gases when stored in air. Absorption of CO_2 was particularly rapid, but CO, N_2, and O_2 were also trapped. Of special importance, the calixarene crystals selectively absorbed CO_2 from a CO_2 - H_2 mixture, leaving the H_2 behind. This phenomenon can be used for purification of H_2. More recently, Atwood showed that calix[4]arene crystals can also absorb H_2 at higher pressures [183]. Cavity-containing solid materials for gas entrapment and storage have thus emerged. While polymers with intrinsic calixarene cavities have not yet been constructed [184], He, H_2, N_2, N_2O, and CO_2 were encapsulated in the solid state by hemicarcerand [185]. These gases were shown to replace each other in the solid hemicarcerand. The scope of gas encapsulation was thus expanded from solution to the gas-solid interface. In summary, calixarenes can be employed in the design of cavity containing solid materials for gas entrapment, storage and release. Polymers with intrinsic calixarene cavities are still not known, but this is just a matter of time. The major drawback of reversible encapsulation complexes with gases is in their low thermodynamic stability. Even well preorganized cavities of calixarenes and hemicarcerands derived from them cannot complex strongly, due to the lack of binding interactions. An alternative approach is based on reversible chemical transformation of gases upon complexation. In this case, they produce reactive intermediates with higher affinities for the receptor molecules.

Chapter 6

TRANSPORT OF SMALL MOBILE PARTICLES THROUGH A CALIXARENE CRYSTAL

More than two decades ago, Andreetti and coworkers reported the X-ray structural analysis of a 1:1 complex of *p*-tert-butylcalix[4]arene with toluene, **1a** [23]. This pioneering study unambiguously confirmed the cone conformation of the host molecule, and showed that its cavity is capable of accommodating small aromatic species. Subsequent studies have revealed that inclusion compounds involving *p*-tert-butylcalix[4]arene most often crystallize with the host molecules packed as bilayers. Furthermore, the bilayer packing mode can be subdivided into two major categories (figure 7): **A**, space group $P4/n$, host:guest ratio 1:1 and the guest is partially inserted into the calixarene cavity; **B**, space group $P4/nnc$, host:guest ratio 2:1 and two calixarene molecules face one another to form a dimeric capsule which can completely enclose the guest.

Thermogravimetric analysis of the 1:1 host:guest complex of *p*-tert-butylcalix[4]arene and toluene yields two distinct weight-loss events, each accounting for half of the total amount of toluene originally present in the material. The first weightloss occurs with an onset temperature of 108 °C. It was shown [15] that this corresponds to a transition from a structure of type **A** to one of type **B** as adjacent bilayers shift laterally by *ca.* 9 Å relative to one another. This process occurs as a single-crystal-to-single-crystal phase transformation and is triggered by relatively weak van der Waals interactions between host and guest molecules. Loss of the remaining toluene occurs with an onset temperature of 120 °C, yielding the host material in its pure, unsolvated form.

Ripmeester reported the crystal structure of pure *p*-tert-butylcalix[4]arene grown over a period of three days at 70 °C from a tetradecane solution of the compound [138]. In this phase, **1b**, the host molecules associate with one another as mutually self-included dimers in which each molecule inserts one its But-groups into the other's cavity. This arrangement is relatively well packed with an efficiency of 0.67. In the absence of any known polymorphic structures of pure *p*-tert-butylcalix[4]arene, it was reasonable to speculate that **1b** might be the phase that obtains upon guest removal. Indeed, this assumption was endorsed in a subsequent report [157] dealing with solid–vapor guest inclusion/decomposition processes involving *p*-tert-butylcalix[4]arene and a related compound. Therefore, in light of the recent interest in solid–solid phase transitions of *p*-tert-butylcalix[4]arene, it is relevant to finally address this issue conclusively.

Atwood obtained diffraction-quality single crystals of *p*-tert-butylcalix[4]arene by sublimation of the compound at 280 °C under reduced pressure [11]. The structure of this sublimed phase, **1c**, has already been described in detail [15]. Phase **1c** proves to be a polymorphic form of pure *p*-tert-butylcalix[4]arene and, in contrast to **1b**, has a relatively low packing efficiency of 0.59. The molecules pack in the familiar bilayer motif with facing calixarenes slightly offset relative to one another (figure 7). This arrangement results in the formation of skewed capsules, each with an estimated free volume of 235 Å^3. Furthermore, these capsules are unoccupied, thus accounting for the rather low efficiency of packing. Following previously reported procedures [23, 138] Atwood also prepared crystals of **1a** and **1b**. The former were heated at 220 °C under reduced pressure for three hours in order to remove all of the toluene guest molecules. X-Ray powder diffraction patterns of the various materials were recorded and it was revealed that, upon removal of toluene by heating, **1a** undergoes a phase transition to eventually yield **1c**.

Atwood and coworkers have similarly investigated 1:1 host:guest adducts of *p*-tert-butylcalix[4]arene with *p*-chlorotoluene, *p*-fluorotoluene, chlorobenzene and fluorobenzene. In each case, they found that guest removal results in phase **1c**. Although the authors concede that it may still be possible to form phase **1b** by means of guest removal, they have not yet observed this to be the case for the limited series of inclusion compounds considered here. It is important to note that, with regard to the arrangement of the calixarene molecules, the fundamental difference between **1c** and the structures of types **A** and **B** is that adjacent bilayers are shifted laterally relative to one another (figure 7).

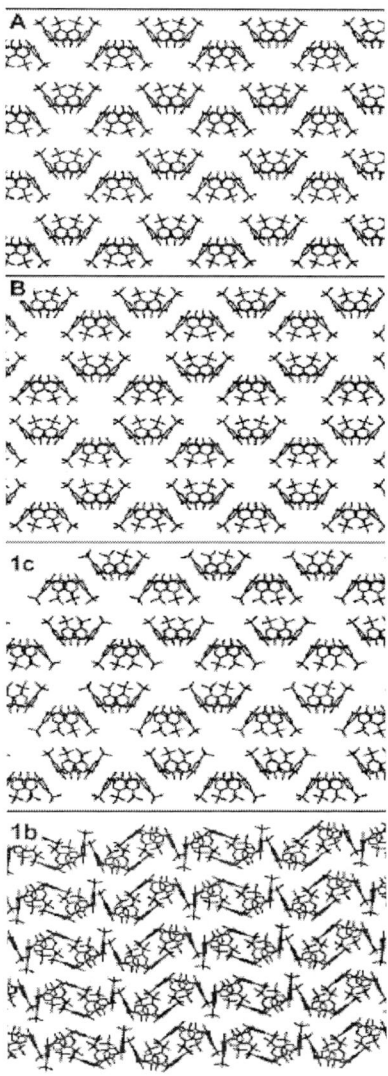

Figure 7. X-Seed projections contrasting 1c with the two most commonly observed bilayer packing arrangements (A and B) of *p*-tert-butylcalix[4]arene. In A, facing calixarenes are offset with respect to one another and the guest molecules (not shown) are partially inserted into their cavities. Adjacent bilayers of B are translated laterally relative to those of A such that facing calixarenes form dimeric capsules. In 1c, facing calixarenes are slightly offset and form skewed, unoccupied capsules. The self-included dimer structure, 1b, obtained from a tetradecane solution of *p*-tert-butylcalix[4]arene is also shown.

In all three structural types, the molecular spacing, both within and between the bilayers, is practically identical. The surfaces of the bilayers are composed of bulky But-groups, interspersed with calixarene cavities and the crevices between adjacent But- moieties. Thus it is not difficult to envision how adjacent bilayers are able to slide over one another in order to facilitate guest uptake or release as a solid/liquid or solid/vapor inclusion reaction. As yet, the influence of weak van der Waals interactions on both the stability as well as the dynamics of molecules in the solid state has been largely overlooked. However, Atwood and coworkers have shown that studies of simple, yet elegant model systems provide valuable insight into the organic solid state, and that a high degree of cooperativity exists between molecules during phase transitions.

Single-crystal to single-crystal phase transformations have received considerable attention in recent years [186]. However, very few examples are known to date because crystals normally do not retain the single crystallanity after the transformation. Most of the reported cases involve guest exchange in porous materials in which structural transformation of the host framework is triggered by guest exchange or removal [187]. Exchange of guest molecules in porous metal-organic frameworks is a well-known but poorly understood phenomenon [188-190]. In contrast, single-crystal to single-crystal transformations caused by guest inclusion in crystals that are held together by weak van der Waal forces are very rare [186]. Host framework retention is a useful strategy for the storage and separation, storage, and transportation of gases [191-195]. Calixarenes are versatile inclusion compounds that have been studied for gas sorption and sensor applications [16, 196-200]. The adsorption of gases, particularly carbon dioxide, is very important for carbon capture and sequestration technology and is one of the most challenging aspects of transportation of carbon dioxide. In this regard, Atwood and coworkers have reported several papers on the CO_2 absorption properties of calix[4]arene derivatives, which demonstrate advantages compared to other porous materials [201]. During the process of gas uptake and release, the host framework should not undergo phase transformation because such transformations are associated with an energy penalty. For instance, Atwood and coworkers have recently reported the "gas-induced transformation and expansion" of an organic solid upon CO_2 or N_2O uptake [202]. The authors discovered that the guest-free thermodynamic form of *p*-tert-butylcalix[4]arene may be converted to the 1:1 host:guest tetragonal form. Such a process should involve a considerable energy barrier, and surprisingly, this transformation process occurs much faster at low temperatures (-10 °C). However, low pressures (100 psi) may require up to 10 days to effect the transformation. In communication [203], Atwood et al. have found that a range of gases as well as solvent molecules can effect the transformation of guest-free high-density *p*-tert-butylcalix[4]arene to the 1:1 host:guest tetragonal form.

Similarly, they extended this phenomenon to a range of gases and solvents molecules using the guest free thermodynamic form under ambient conditions. At room temperature and 0.70 MPa (100 psig) of propane, **1b** was found to transform to **A** in just 10 min. The presence of indicative peaks corresponding to phase of 1:1 host:guest form in the powder diffaction pattern suggests that the transformation from **1b** to **A** occurs through phase of *p*-tert-butylcalix[4]arene complex. As a result of the transformation, the crystal volume increases by 13% going from **1b** to **A**. The powder plot of **A** with propane inclusion is near-identical to that of **A** with CO_2 and N_2O. However, the transformation from **1b** to **A** with CO_2 is rather slow and it is possible to isolate the kinetic form *p*-tert-butylcalix[4]arene with CO_2 (i.e., as a 2:1 host:guest phase **B** prior to conversion to the 1:1 form **A** over a given period of time). As a result, **1b** expands 33% along the [100] direction and shrinks by 17 and 7% along the the [010] and [001] directions, respectively, upon changing to *p*-tert-butylcalix[4]arene complex. The experimental powder plots with these gases are near-identical to that of simulated powder patterns of the 1:1 host:guest complex suggest these gases can form 1:1 host:guest inclusion complexes with *p*-tert-butylcalix[4]arene. Because of the near-identical powder patterns, it is possible to speculate that each calixarene void is filled with one gas molecule. Several experiments were conducted to determine the exact loacation of the gas molecules in the 1:1 complex using single-crystal X-ray diffraction experiments, but all failed. Though the powder looked identical to the 1:1 host:guest complex, it's possible that the location of gas molecules may look different. To further confirm the molar ratios of the various gases in *p*-tert-butylcalix[4]arene, gas sorption measurements were conducted on **1b** at various pressures of CO_2, N_2O, and propane using volumetric analyzer. Previous studies on *p*-tert-butylcalix[4]arene complex at low pressures of CO_2 and N_2O (moles of gas:calixarene ratio = 0.5:1) showed that only a single molecule of the gas can be accommodated within each void formed by two calixarene molecules [204-206], i.e., host:guest = 2:1. Similar experiments on **1b** at 1 atm pressure indicate phase **1b** is dead for sorption, which further supports the observation in which at lower pressures (0.70 MPa, 100 psig) of CO_2 and N_2O the conversion from **1b** to **A** is very slow and takes as long as 10 days to a month. However, the same measurements on *p*-tert-butylcalix[4]arene complex and **1b** at high pressure indicate instantaneous pressure drop and reaches equilibrium in just 2 h. At this pressure, 1.07 moles of gas occupied per calixarene void, i.e., host:guest = 1:1. Similar experiments on **1b** at various pressures of propane indicate the molar ratio of the propane approaches close to one. At 1, 3, and 7 bar of propane, the molar ratio is found to be 0.3, 0.5, and 0.7. Therefore, it is possible to selectively transform **1b** to the desired host:guest = 1:1, 2:1, or 1:2 forms simply through gas selection or by employing higher pressures of the same gas.

The authors also examined the effects of solvents and their vapors on form **1b**. Freshly grown crystals of **1b** were placed in two separate vails. One set of crystals was soaked in acetone, whereas for the others, CS_2 vapors were allowed to diffuse through the crystals for 10 min. Both crystal batches were crushed into fine powders and the corresponding powder plots indicate the transformation from **1b** to the 1:1 host:guest (acetone and CS_2) forms. Shatz et al. [207] reported the formation of an acetone complex with a 2:1 host:guest ratio, which unfortunately has not been fully characterized. The powder patterns suggest that the molar ratio of host and guest is 1:1. The observations are in good agreement with the simulated powder diffraction plots of 1:1 acetone and CS_2 complexes obtained simply by recrystallizing *p*-tert-butylcalix[4]arene from CS_2 and acetone, as reported by Klinoski and Schatz et al. [207, 208]. X-ray analysis on **1b** at various temperatures suggests that at low temperature, three out of four *tert*-butyl groups are disordered over two positions, whereas all are disordered at room temperature. Therefore, one can further speculate that limited rotation of *tert*-butyl groups in **1b** would allow the gas molecules to diffuse through the solid, and simultaneously the energy released after the gas is trapped in crystal is sufficient to transform **1b** to the more stable form **A** or **B** with included gas or solvent molecules. The existence of 2:1, 1:1, and higher gas inclusion complexes and failure of similar transformation to occur with other gases (H_2, He, and CH_4) suggest the possibility of exploiting these materials for carbon capture and separation applications.

REFERENCES

[1] Nabok, A.V.; Hassan, A.K.; Ray, A.K. *J. Mater. Chem.* 2002, *10*, 189-194.
[2] Chaabane, R.B.; Gamoudi, M.; Guillaud, G.; Jouve, C.; Lamartine, R.; Bouazizi, A.; Maaref, H. *Sens. Actuators B* 1996, *31*, 41-44.
[3] Mlika, R.; Dumazet, I.; Gamoudi, M.; Lamartine, R.; Ben Ouada, H.; Jaffrezic-Renault, N.; Guillaud, G. *Anal. Chem. Acta* 1997, *354*, 283-289.
[4] Mlika, R.; Ben Ouada, H.; Hamza, M.A.; Gamoudi, M.; Guillaud, G.; Jaffrezic-Renault, N. *Synth. Met.*. 1997, *90*, 173-179.
[5] Mlika, R.; Ben Ouada, H.; Jaffrezic-Renault, N.; Dumazet, I.; Lamartine, R.; Gamoudi, M.; Guillaud, G. *Sens. Actuators B* 1998, *47*, 43-47.
[6] Abraham, W. *J. Incl. Phenom. Macrocycl. Chem.* 2002, *43*, 159-174.
[7] Hirakata, M.; Yoshimura, K.; Usui, S.; Nishimoto K.; Fukazawa, Y. *Tetrahedron Lett.* 2002, *43*, 1859-1861.
[8] Vincenti, M. *J. Mass Spectrosc.* 1995, *30*, 925-939.
[9] Inokuchi, F.; Araki, K.; Shinkai, S. *Chem. Lett.* 1994, *23(8)*, 1383-1386.
[10] Vincenti, M.; Irico, A. *Int. J. Mass Spectrom.* 2002, *214*, 23-36.
[11] Atwood, J. L.; Barbour, L. J.; Jerga, A. *Chem. Commun.* 2002, 2952-2953.
[12] Atwood, J. L.; Barbour, L. J.; Lloyd, G. O.; Thallapally, P. K. *Chem. Commun.* 2004, 922-923.
[13] Surov, O. V.; Mamardashvili, N. Zh.; Shaposhnikov, G. P.; Koifman, O. I. *Russ. J. Gen. Chem.* 2006, *76 (6)*, 974-979.
[14] Surov, O. V.; Mamardashvili, N. Zh.; Shaposhnikov, G. P.; Koifman, O. I. *Russ. J. Phys. Chem. A* 2007, *81(12)*, 1936-1941.

[15] Atwood, J. L.; Barbour, L. J.; Jerga, A.; Schottel, B. L. *Science* 2002, *298*, 1000-1002.
[16] Thallapally, P. K.; Lloyd, G. O.; Atwood, J. L.; Barbour, L. J. *Angew. Chem. Int. Ed.* 2005, *44 (25)*, 3848-3851.
[17] Alcaciz-Monge, A.; Linares-Solano, A.; Rand, B. *J. Phys. Chem. B* 2001, *105*, 7998-8001.
[18] Gutache, C. D.; Dhawan, B.; No, K. H.; Muthukrishnan, R *J. Am. Chem. Soc.* 1981, *103*, 3782-3791.
[19] Coruzzi, M.; Andreetti, G. D.; Bocchi, V.; Pochini, A.; Ungaro, R. *J. Chem. Soc. Perkin Trans. 2* 1982, *11*, 33-44.
[20] Ninagawa, A.; Matauda, H. *Makromol. Chem. Rapid Commun.* 1982, *3*, 65-67.
[21] Coruzzi, M.; Andreetti, G. D.; Bocchi, V.; Pochini, A.; Ungaro, R. *J. Chem. Soc. Perkin Trans. 2 1982*, 1133-1141.
[22] Nakamoto, Y.; Ishida, S. *Makromol. Chem. Rapid Commun.* 1982, *3*, 705-706.
[23] Andreetti, G. D.; Ungaro, R.; Pochini, A. *J.* Chem. *Soc. Chem. Commun.* 1979, 1005-1007.
[24] Gutsche, C.D. *Acc. Chem. Res.* 1983, *16*, 161-170.
[25] Gutsche, C. D. In *Calixarenes Revisited;* Stoddart, J.F.; Ed.; Monographs in Supramolecular Chemistry; The Royal Society of Chemistry: Cambridge, U.K., 1998; Vol. 6.
[26] Pochini, A.; Ungaro R. In *Comprehensive Supramolecular Chemistry*; Vögtle, F.; Ed.; Elsevier: 1996; Vol. 2, Chapt. 4, pp 103-142.
[27] Nishio, M.; Hirota, M.; Umezawa Y. In *The CH/π Interaction;* Marchand, A.P; Ed.; Methods in Stereochemical Analysis; Wiley-VCH: 1998; No. 11.
[28] Umezawa, Y.; Tsuboyama, S.; Honda, K.; Uzawa, J.; Nishio, M. *Bull. Chem. Soc. Jpn.* 1998, *71*, 1207-1213.
[29] Braga, D.; Grepioni, F.; Tedesco, E. *Organometallics* 1998, *17*, 2669-2672.
[30] Arduini, A; Pochini, A; Secchi, A; Ugozzoli, F. In *Calixarenes 2001;* Asfari, Z. et al.; Eds.; Kluwer Academic Publishers: Dordrecht, NL, 2001; Chapt. 25, pp 457–475.
[31] Brouwer, E.B.; Enright, G.D.; Ratcliffe, C.I.; Ripmeester, J.A.; Udachin, K.A. In *Calixarenes 2001;* Asfari, Z. et al.; Eds.; Kluwer Academic Publishers: Dordrecht, NL, 2001; Chapt. 16, pp 296-311.
[32] Arduini, A.; McGregor, W. M.; Paganuzzi, D.; Pochini, A.; Secchi, A.; Ugozzoli, F.; Ungaro, R. *J. Chem.Soc. Perkin Trans. 2* 1996, 839-846.

[33] Abidi, R.; Baker, M. V.; Harrowfield, J. M.; Ho, D. S.-C.; Richmond, W. R.; Skelton, B. W.; White, A. H.; Varnek, A.; Wipff, G. *Inorg. Chim. Acta* 1996, *246*, 275-286.
[34] McKervey, M. A.; Seward, E. M.; Ferguson, G.; Ruhl, B. L. *J. Org. Chem.* 1986, *51*, 3581-3584.
[35] Gibson, V. C.; Redshaw, C.; Clegg, W.; Elsegood, M. R. *J. J. Chem. Soc. Chem. Commun.* 1997, 1605-1606.
[36] Harrowfield, J. M.; Ogden, M. I.; Richmond, W. R.; Skelton, B. W.; White, A. H. *J. Chem. Soc. Perkin Trans.2* 1993, 2183-2190.
[37] Beer, P. D.; Drew, M. G. B.; Leeson, P. B.; Lyssenko, K.; Ogden, M. I. *J. Chem. Soc. Chem. Commun.* 1995, 929-930.
[38] Beer, P.D.; Drew, M. G. B.; Leeson, P. B.; Ogden, M. I. *J. Chem. Soc. Dalton Trans.* 1995, 1273-1283.
[39] Xu, W.; Puddephatt, R. J.; Manojlovic-Muir, L.; Muir, K. W.; Frampton, C. S. *J. Incl. Phenom. Macrocycl. Chem.* 1994, *19*, 277-290.
[40] MacGillivray, L. R.; Atwood, J. L. *J. Am. Chem. Soc.* 1997, *119*, 6931-6932.
[41] Zhen-Lin, Z.; Yuan-Yin, C.; Xue-Ran, L.; Bao-Sheng, L.; Liao-Rong, C. *Jiegou Huaxue (J. Struct. Chem.).* 1996, *15*, 358-360.
[42] Reichwein, A. M.; Verboom, W.; Harkema, S.; Spek, A. L.; Reinhoudt, D. N. *J. Chem. Soc. Perkin Trans.2* 1994, 1167-1172.
[43] Aleksiuk, O.; Grynszpan, F.; Biali, S. E. *J. Incl. Phenom. Macrocycl. Chem.* 1994, *19*, 237-256.
[44] Biali, S. E.; Böhmer, V.; Cohen, S.; Ferguson, G.; Güttner, C.; Grynszpan, F.; Paulus, E. F.; Thondorf, I.;Vögt, W. *J. Am. Chem. Soc.* 1996, *118*, 12938-12949.
[45] Szemes, F.; Hesek, D.; Chen, Z.; Dent, S. W.; Drew, M. G. B.; Goulden, A. J.; Graydon, A. R.; Grieve, A.; Mortimer, R. J.; Wear, T.; Weightman, J. S.; Beer, P. D. *Inorg.Chem.* 1996, *35*, 5868-5879.
[46] Danil de Namor, A. F.; Piro, O. E.; Pulcha Salazar, L. E.; Aguilar-Cornejo, A. F.; Al-Rawi, N.; Castellano, E. E.; Sueros Velarde, F. J. *J. Chem. Soc. Faraday Trans.* 1998, *94*, 3097-3104.
[47] Verboom, W.; Struck, O.; Reinhoudt, D. N.; van Duynhoven, J. P. M.; van Hummel, G. J.; Harkema, S.; Udachin, K.A.; Ripmeester, J. A. *Gazz. Chim. Ital.* 1997, *127*, 727-739.
[48] Pitarch, M.; Walker, A.; Malone, J. F.; McGarvey, J .J.; McKervey, M. A.; Creaven, B.; Tobin, D. *Gazz.Chim. Ital.* 1997, *127*, 717-721.
[49] Akdas, H.; Bringel, L.; Graf, E.; Hosseini, M. H.; Mislin, G.; Pansanel, J.; De Cian, A.; Fischer, J. *Tetrahedron Lett.* 1998, *39*, 2311.

[50] Airola, K.; Böhmer, V.; Paulus, E. F.; Rissanen, K.; Schmidt, C.; Thondorf, I.; Vogt, W. *Tetrahedron* 1997, *53*, 10709-10724.
[51] Giannini, L.; Caselli, A.; Solari, E.; Floriani, C.; Chiesi-Villa, A.; Rizzoli, C.; Re, N.; Sgamellotti, A. *J. Am.Chem. Soc.* 1997, *119*, 9709-9719.
[52] Beer, P. D.; Drew, M. G. B.; Grieve, A.; Kan, M.; Leeson, P. B.; Nicholson, G.; Ogden, M. I.; Williams, G. *Chem. Commun.* 1996, 1117-1118.
[53] Gardiner, M. G.; Koutsantonis, G. A.; Lawrence, S. M.; Nichols, P. J.; Raston, C. L. *Chem. Commun.* 1996, 2035-2036.
[54] Shevchenko, I.; Zhang, H.; Lattman, M. *Inorg. Chem.* 1995, *34*, 5405-5409.
[55] Beer, P.D.; Drew, M. G. B.; Grieve, A.; Ogden, M. I. *J. Chem. Soc. Dalton Trans.* 1995, 3455-3466.
[56] Beer, P.D.; Drew, M. G. B.; Kan, M.; Leeson, P. B.; Ogden, M. I.; Williams, G. *Inorg. Chem.* 1996, *35*, 2202-2211.
[57] Beer, P. D.; Drew, M. G. B.; Leeson, P. B.; Ogden, M. I. *Inorg. Chim. Acta.* 1996, *246*, 133-141.
[58] Mislin, G.; Graf, E.; Hosseini, M. H.; Bilyk, A.; Hall, A. K.; Harrowfield, J. M.; Skelton, B. W.; White, H. *Chem. Commun.* 1999, 373-374.
[59] Böhmer, V.; Ferguson, G.; Gallagher, J. F.; Lough, A. J.; McKervey, M. A.; Madigan, E.; Moran, M. B.; Phillips, J.; Williams, G. *J. Chem. Soc. Perkin Trans.1* 1993, 1521-1527.
[60] Leigh, D. A.; Linnane, P.; Pritchard, R. G.; Jackson, G. *Chem. Commun.* 1994, 389-390.
[61] Helgeson, R. C.; Knobler, C .B.; Cram, D. J. *Chem. Commun.* 1995, 307-308.
[62] Sartori, G.; Porta, C.; Bigi, F.; Maggi, R.; Peri, F.; Marzi E. *Tetrahedron* 1997, *53*, 3287-3300.
[63] Thuéry, P.; Nierlich, M.; Lamare, V.; Dozol, J.-F.; Asfari, Z.; Vicens, J. *J. Incl. Phenom. Macrocycl. Chem.* 2000, *36*, 375-408.
[64] Bauer, L.J.; Gutsche, C.D. *J. Am. Chem. Soc.* 1985, *107*, 6063-6069.
[65] Gutsche, C. D.; Levine, J. A. *J. Am. Chem. Soc.* 1982, *104*, 2652-2659.
[66] Danil de Namor, A.F.; Zapata-Ormachea, M.L.; Hutcherson, R.G. *J. Phys. Chem. B* 1998, *102*, 7839-7851.
[67] Danil de Namor, A.F.; Hutcherson, R.G.; Sueros Velarde, J.F.; Alvarez-Larena, A.; Briansó-Penalva, J.L. *J. Chem. Soc. Perkin Trans. 1* 1998, 2933-2940.

[68] Danil de Namor, A. F.; Zapata-Ormachea, M.L.; Hutcherson, R.G. *J. Phys. Chem. B* 1999, *103*, 366-377.
[69] Arena, G.; Contino, A.; Gulino, F. G.; Magr´ı, A.; Sciotto, D.; Ungaro, R. *Tetrahedron Lett.* 2000, *41*, 9327-9336.
[70] Kunsági-Máté, S.; Nagy, G.; Kollár, L. *Anal. Chim. Acta* 2001, *428*, 301-308.
[71] Kunsági-Máté, S.; Nagy, G.; Kollár, L. *Sens. Actuators B* 2001, *76*, 545-549.
[72] Kunsági-Máté, S.; Bitter, I.; Grün, A.; Nagy, G.; Kollár, L. *Anal. Chem. Acta* 2002, *461*, 273–279.
[73] Danil de Namor, A. F.; Cleverley, R. M.; Zapata-Ormachea, M. L. *Chem. Rev.* 1998, *98*, 2495-2525.
[74] Danil de Namor, A.F. *Pure Appl. Chem.* 1993, *65*, 193-197.
[75] Danil de Namor, A.F.; Hutcherson, R.G.; Sueros Velarde, F.J.; Zapata-Ormachea, M.L.; Pulcha Salazar, L.E.; al Jammaz, I.; al Rawi N. *Pure Appl. Chem.* 1998, *70*, 769-774.
[76] Danil de Namor, A.F.; Cleverley, R.R.; Zapata-Ormachea M.L. *Chem. Rev.* 1998, *98*, 2495-2497.
[77] Danil de Namor, A.F.; Kowalska, D.; Marcus, Y.; Villanueva-Salas, J. *J. Phys. Chem. B* 2001, *105*, 7542-7548.
[78] Danil de Namor, A.F.; Kowalska, D.; Castellano, E.E.; Piro, O.E.; Sueros Velarde, F.J.; Villanueva-Salas, J. *Phys. Chem. Chem. Phys.* 2001, *3*, 4010-4016.
[79] Danil de Namor, A.F.; Chahine, S.; Castellano, E.E.; Piro, O.E. *J. Phys. Chem B* 2004, *108*, 11384-11394.
[80] Danil de Namor, A.F.; Aguilar-Cornejo, A.; Soualhi M. Shehab, R.; Nolan, K.B.; Ouazzani, N.; Mand, L. *J. Phys. Chem. B* 2005, *109*, 14735-14741.
[81] Liu, Y.; Wang, H.; Wang, L.-H.; Zhang, H.-Y. *Thermochim. Acta* 2004, *414*, 65-71.
[82] Liu, Y.; Yang, E.-C.; Chen, Y. *Thermochim. Acta* 2005, *429*, 163-171.
[83] Liu, Y.; Yang, E.C.; Chen, Y.; Guo, D.S.; Ding, F. *Eur. J. Org. Chem.* 2005, *21*, 4581-4585.
[84] Vicens, J.; Böhmer, V. *Calixarenes: A Versatile Class of Macrocyclic Compounds*; Kluver Academic Publishers: Dordrecht, NL, 1991.
[85] Vicens, J.; Asfari, Z.; Harrowfield, J.M. *Calixarenes 50th Anniversary*; Commemorative Volume; Kluver Academic Publishers: Dordrecht, NL, 1994.
[86] Böhmer, V. *Angew. Chem. Int. Ed.* 1995, *34*, 713-718.

[87] Ikeda, A.; Shinkai, S. *Chem. Rev.* 1997, *97*, 1713-1717.
[88] van Wageningen, A.M.A.; Verboom, W.; Reinhoudt, D.N. *Pure Appl. Chem.* 1996, *68,* 1273-1276.
[89] Mandorlini, L. In *Calixarenes in Action*; Ungaro, R.; Ed.; Imperial College Press: London, U.K., 2000.
[90] Brodbelt, J.S.; Dearden, D.V. In *Mass Spectrometry, Comprehensive Supramolecular Chemistry*; Davies, J.E.D.; Ripmeester, J.A.; Eds.; Pergamon Press: Oxford, U.K., 1996; Vol. 8.
[91] Sawada, M. *Mass Spectrom. Rev.* 1997, *16*, 73-84.
[92] Schalley, C.A. *Int. J. Mass Spectrom.* 2000, *194,* 11-23.
[93] Loo, J.A. *Mass Spectrom. Rev.* 1997, *16*, 1-20.
[94] Wong, P.S.H.; Yu, X.J.; Dearden, D.V. *Inorg. Chim. Acta* 1996, *246,* 259-268.
[95] Schalley, C.A.; Castellano, R.K.; Brody, M.S.; Rudkevich, D.M.; Siuzdak, G.; Rebek Jr., J. *J. Am. Chem. Soc.* 1999, *121*, 4568-4579.
[96] Brody, M.S.; Schalley, C.A.; Rudkevich, D.M.; Rebek Jr., J. *Angew. Chem. Int. Ed.* 1999, *38*, 1640-1642.
[97] Jolliffe, K.A.; Crego Calama, M.; Fokkens, R.; Nibbering, N.M.M.; Timmerman, P.; Reinhoudt, D.N. *Angew. Chem. Int. Ed.* 1998, *37*, 1247-2249.
[98] Cardullo, F.; Crego Calama, M.; Snellink-Ruël, B.H.M.; Weidmann, J.-L.; Bielejewska, A.; Fokkens, R.; Nibbering, N.M.M.; Timmerman, P.; Reinhoudt, D.N. *J. Chem. Soc. Chem. Comm.* 2000, 367-369.
[99] Vincenti, M.; Irico, A. *Int. J. Mass Spectrom.* 2002, *214*, 23-36.
[100] Vincenti, M.; Minero, C.; Pelizzetti, E.; Secchi, A.; Dalcanale, E. *Pure Appl. Chem.* 1995, *67*, 1075-1079.
[101] Liang, T.-M.; Laali, K.K.; Cordero, M.; Wesdemiotis, C. *J. Chem. Research (S)*1991, 354-355.
[102] Vincenti, M.; Dalcanale, E.; Soncini, P.; Guglielmetti, G. *J. Am. Chem. Soc.* 1990, *112*, 445-451.
[103] Cotter, R.J. *Anal. Chem.* 1980, *52*, 1589A.
[104] Surov, O.V.; Mamardashvili, N.Zh.; Shaposhnikov, G.P.; Koifman, O.I. *J. Incl. Phenom. Macrocycl. Chem.* 2007, *58*, 329-335.
[105] Ghidini, E.; Ugozzoli, F.; Ungaro, R.; Harkema, S.; Abu El-Fadl, A.; Reinhoudt, D.N. *J. Am. Chem. Soc.* 1990, *112(19)*, 6979-6985.
[106] Klenke, B.; Friederichsen, W. *J. Chem. Soc. Perkin Trans. 1* 1998, *998(20)*, 3377-3380.
[107] Yam, V.W.W.; Cheung, K.L.; Yuan, L.H.; Wong, K.M.C.; Cheung, K.K. *Chem. Commun.* 2000, *16*, 1513-1514.

[108] van Loon, J-D.; Arduini, A.; Coppi, L.; Verboom, W.; Pochini, A.; Ungaro, R.; Harkema, S.; Reinhoudt, D. *J. Org. Chem.* 1990, *55*, 5639-5646.
[109] Chickos, J.S.; Acree, Jr, W.E. *J. Phys. Chem. Ref. Data* 2002, *31(2)*, 537-698.
[110] Grootenhuis, P.D.J.; Kollman, P.A.; Groenen, L.C.; Reinhoudt, D.N.; van Hummel, G.J.; Ugozzooli, F.; Andreetti, G.D. *J. Am. Chem. Soc.* 1990, *112(11)*, 4165-4176.
[111] Kunugi, Y.; Nigorikawa, K.; Harima, Y.; Yamashita, K. *J. Chem. Soc. Chem. Commun.* 1994, 873-874.
[112] Lu, C.J.; Shin, J.S. *Anal. Chem. Acta* 1995, *306*, 129-133.
[113] Chen, Z.K.; Ng, S.C.; Li, S.F.Y.; Zhong, L.; Xu, L.G.; Chan, H.S.O. *Synth. Met.* 1997, *87*, 201-211.
[114] Zhou, X.C.; Zhong, L.; Li, S.F.Y.; Ng, S.C.; Chan, H.S.O. *Sens. Actuators B* 1997, *42*, 59-68.
[115] Barko, G.; Hlavay, J. *Anal. Chim. Acta* 1998, *367*, 135-142.
[116] Deng, Z.P.; Stone, D.C.; Thompson, M. *Analyst* 1996, *121*, 1341-1349.
[117] Cao, Z.; Cao, D.; Lei, Z.G.; Lin, H.G.; Yu, R.Q. *Talanta* 1997, *44*, 1413-1419.
[118] Papes, V.; Brodska, S. *Sens. Actuators B* 1997, *40*, 143-150.
[119] Costello, B.P.J.D.; Evans, P.; Ewen, R.J.; Honeybourne, C.L.; Ratcliffe, N.M. *J. Mater. Chem.* 1996, *6*, 289-297.
[120] Milella, E.; Musio, F.; Alba, M.B. *Thin Solid Films* 1996, *285*, 908-911.
[121] De Wit, M.; Vanneste, E.; Geise, H.J.; Nagels, L.J. *Sens. Actuators B* 1998, *50*, 164-166.
[122] Buhlmann, K.; Schladtt, B.; Cammann, K.; Shulga, A. *Sens. Actuators B* 1998, *49*, 156-159.
[123] Dickert, F.L.; Baumler, U.P.A.; Zwissler, G.K. *Synth. Met.* 1993, *61*, 47-53.
[124] Nelli, P.; Delcanale, E.; Faglia, G.; Sberveglieri, G.; Soncini, P. *Sens. Actuators B* 1993, *13-14*, 302-312.
[125] Schierbaum, K.D.; Weiss, T.; Thoden van Velzen, E.U.; Engbersen, J.F.J.; Reinhoudt, D.N.; Gopel, W. *Science* 1994, *265*, 1413-1415.
[126] Dalcanale, E.; Hartman, J. *Sens. Actuators B* 1995, *24*, 39-41.
[127] Rickert, J.; Weiss, T.; Gopel, W. *Sens. Actuators B* 1996, *31*, 45-50.
[128] Nabok, A.V.; Lavrik, N.V.; Kazantseva, Z.I.; Nesterenko, B.A.; Markovskiy, L.N.; Kalchenko, V.I.; Shivaniuk, A.N. *Thin Solid Films* 1995, *259*, 244-248.

[129] Nabok, A.V.; Hassan, A.K.; Ray, A.K.; Omar, O.; Kalchenko, V.T. *Sens. Actuators B* 1997, *45*, 115-118.
[130] Nabok, A.V.; Hassan, A.K.; Ray, A.K. *J. Mater. Chem.* 2000, *10*, 189-194.
[131] Schierbaum, K.-D.; Gerlach, A.; Göpel, W.; Müller, W. M.; Vögtle, F.; Dominik, A.; Roth, H. *J. Fresenius J. Anal. Chem.* 1994, *349*, 372-379.
[132] Abraham, M. H.; Platts, J. A. *J. Org. Chem.* 2001, *66*, 3484-3496.
[133] Lehn, J.-M. *Supramolecular Chemistry: Concepts and Perspectives*; VCH: Weinheim, BD, 1995.
[134] Weber, E. In *Comprehensive Supramolecular Chemistry*; Atwood, J. L., Davies, J. E., MacNicol, D. D., Vogtle, F., Eds.; Elsevier Science: Oxford, U.K., 1996; Vol. 6, Chapt. 17, pp 535-592.
[135] Weber, E.; Hens, T.; Brehmer, T.; Csöregh, I. *J. Chem. Soc. Perkin Trans. 2* 2000, 235-238.
[136] Gorbatchuk, V. V.; Tsifarkin, A. G.; Antipin, I. S.; Solomonov, B. N.; Konovalov, A. I. *Mendeleev Commun.* 1999, 11-13.
[137] Gorbatchuk, V. V.; Tsifarkin, A. G.; Antipin, I. S.; Solomonov, B. N.; Konovalov, A. I.; Seidel, J.; Baitalov, F. *J. Chem. Soc. Perkin Trans. 2* 2000, 2287-2289.
[138] Brouwer, E. B.; Udachin, K. A.; Enright, G. D.; Ripmeester, J. A.; Ooms, K. J.; Halchuk, P. A. *Chem. Commun.* 2001, 565-567.
[139] Fredeslund, A.; Jones, R. L.; Prausnitz, J. M. *AIChE J.* 1975, *21*, 1086-1089.
[140] Bishop, R. *Chem. Soc. Rev.* 1996, *25*, 311-313.
[141] Weber, E.; Wierig, A.; Scobridis, K. *J. Prakt. Chem.* 1996, *338*, 553-558.
[142] Weber, E.; Hens, T.; Gallardo, O.; Csöregh, I. *J. Chem. Soc. Perkin Trans. 2* 1996, 737-742.
[143] Weber, E.; Hens, T.; Li, Q.; Mak, T. C. W. *Eur. J. Org. Chem.* 1999, 1115-1121.
[144] Schatz, J.; Schildbach, F.; Lentz, A.; Rastätter, S. *J. Chem. Soc. Perkin Trans. 2* 1998, 75-82.
[145] Dickert, F. L.; Keppler, M.; Zwissler, G. K.; Obermeier, E. *Ber. Bunsen-Ges. Phys. Chem.* 1996, *100*, 1312-1319.
[146] Furusho, Y.; Aida, T. *Chem. Commun.* 1997, 2205-2207.
[147] Coetzee, A.; Nassimbeni, L. R.; Achleitner, K. *Thermochim. Acta* 1997, *298*, 81-88.
[148] Dewa, T.; Endo, K.; Aoyama, Y. *J. Am. Chem. Soc.* 1998, *120*, 8933-8940.

[149] Beketov, K.; Weber, E.; Seidel, J.; Köhnke, K.; Makhkamov, K.; Ibragimov, B. *Chem. Commun.* 1999, 91-93.
[150] Nassimbeni, L. R. In *Molecular Recognition and Inclusion*; Coleman, A. W., Ed.; Kluwer: Dordrecht, NL, 1998; pp 135- 152.
[151] Brouwer, E. B.; Udachin, K. A.; Enright, G. D.; Ripmeester, J. A. *Chem. Commun.* 2000, 1905-1907.
[152] Ungaro, R.; Pochini, A.; Andreetti, G. D.; Sangermano, V. *J. Chem. Soc. Perkin Trans. 2* 1984, 1979-1985.
[153] Toda, F.; Tanaka, K.; Miyahara, I.; Akutsu, S.; Hirotsu, K. *Chem. Commun.* 1994, 1795-1797.
[154] Lee, S.-O.; Harris, K. D. M. *Chem. Phys. Lett.* 1999, *307*, 327-333.
[155] Kuruma, K.; Nakagawa, H.; Imakubo, T.; Kobayashi, K. *Bull. Chem. Soc. Jpn.* 1999, *72*, 1395-1401.
[156] Brehmer, T. H.; Weber, E.; Cano, F. H. *J. Phys. Org. Chem.* 2000, *13*, 63-72.
[157] Gorbatchuk, V. V.; Tsifarkin, A. G.; Antipin, I. S.; Solomonov, B. N.; Konovalov, A. I.; Lhotak, P.; Stibor I. *J. Phys. Chem. B* 2002, *106*, 5845-5851.
[158] Gorbatchuk, V. V.; Tsifarkin, A.G.; Antipin, I. S.; Solomonov, B. N.; Konovalov, A. I. *J. Incl. Phenom. Macrocycl. Chem.* 1999, *35*, 389–396.
[159] [159] Ziganshin, M. A.; Yakimov, A. V.; Konovalov, A. I.; Antipin, I. S.; Gorbatchuk, V. V. *Russ.Chem.Bull. Int.Ed.* 2004, *53(7)*, 1536—1543.
[160] Surov, O. V.; Voronova, M. I. *Russ. J. Phys. Chem. A* 2009, *83(5)*, 822–825.
[161] Jura, G.; Harkins, W. D. *J. Am. Chem. Soc.* 1944, *66(8)*, 1356-1373.
[162] Lhotak, P.; Himl, M.; Stibor, I. et al. *Tetrahedron* 2003, *59*, 7581-7586.
[163] Hirakata, M.; Yoshimura, K.; Usui, S.; Nishimoto, K.; Fukazawa, Y. *Tetrahedron Lett.* 2002, *43*, 1859-1861.
[164] Iwamoto, H.; Hirakata, M.; Usui, S.; Haino, T.; Fukazawa, Y. *Tetrahedron Lett.* 2002, *43*, 85–87.
[165] Endo, K.; Kondo, Y.; Aoyama Y.; Hamada, F. *Tetrahedron Lett.* 2003, *44*, 1355–1358.
[166] Murayama, K.; Aoki, K. *Inorg. Chim. Acta* 1998, *281*, 36–42.
[167] Leverd, P. C.; Berthault, P.; Lance, M.; Nierlich, M. *Eur. J. Inorg. Chem.* 1998, 1859–1862.
[168] Schatz, J.; Schildbach, F.; Lentz, A.; Rastätter, S. *.J. Chem. Soc. Perkin Trans. 2* 1998, 133-144.
[169] Graham, B.F.; Harrowfield, J.M.; Tengrove, R.D.; Lagalante, A.F.; Bruno, T.J. *J. Incl. Phenom. Macrocycl. Chem.* 2002, *43*, 179–182.

[170] James, S. L. *Chem. Soc. Rev.* 2003, *32*, 276-288.
[171] Janiak, C. *J. Chem. Soc. Dalton Trans.* 2003, 2781-2804.
[172] Rosi, N. L.; Eckert, J.; Eddaoudi, M.; Vodak, D. T.; Kim, J.; O'Keeffe, M.; Yaghi, O. M. *Science* 2003, *300*, 1127-1130.
[173] Pan, L.; Sander, M. B.; Huang, X.; Li, J.; Smith, M.; Bittner, E.; Bockrath, B.; Johnson, J. K. *J. Am. Chem. Soc.* 2004, *126*, 1308-1309.
[174] Düren, T.; Sarkisov, L.; Yaghi, O. M.; Snurr, R. Q. *Langmuir* 2004, *20*, 2683-2689.
[175] Eddaoudi, M.; Kim, J.; Rosi, N.; Vodak, D.; Wachter, J.; O'Keeffe, M.; Yaghi, O. M. *Science* 2002, *295*, 469-472.
[176] Kitagawa, S.; Kitaura, R.; Noro, S. *Angew. Chem. Int. Ed.* 2004, *43*, 2334-2375.
[177] Sharma, A. C.; Borovik, A. S. *J. Am. Chem. Soc.* 2000, *122*, 8946-8955.
[178] Padden, K. M.; Krebs, J. F.; MacBeth, C. E.; Scarrow, R. C.; Borovik, A. S. *J. Am. Chem. Soc.* 2001, *123*, 1072-1079.
[179] Rowsell, J. L. C.; Yaghi, O. M. *Angew. Chem. Int. Ed.* 2005, *44*, 4670-4679.
[180] Atwood, J. L.; Barbour, L. J.; Jerga, A. *Science* 2002, 296, 2367-2369.
[181] Enright, G. D.; Udachin, K. A.; Moudrakovski, I. L.; Ripmeester, J. A. *J. Am. Chem. Soc.* 2003, *125*, 9896-9897.
[182] Atwood, J. L.; Barbour, L. J.; Jerga, A. *Angew. Chem. Int. Ed.* 2004, *43*, 2948-2950.
[183] Thallapally, P. K.; Lloyd, G. O.; Wirsig, T. B.; Bredenkamp, M. W.; Atwood, J. L.; Barbour, L. J. *Chem. Commun.* 2005, 5272-5274.
[184] McKeown, N. B.; Budd, P. M.; Msayib, K. J.; Ghanem, B. S.; Kingston, H. J.; Tattershall, C. E.; Makhseed, S.; Reynolds, K. J.; Fritsch, D. *Chem. Eur. J.* 2005, *11*, 2610-2620.
[185] Leontiev, A. V.; Rudkevich, D. M. *Chem. Commun.* 2004, 1468-1469.
[186] Atwood, J. L.; Barbour, L. J.; Jerga, A. *Science* 2006, *296*, 2367-2369.
[187] Kitagawa, S.; Kitaura, R.; Noro, S.-I. *Angew. Chem. Int. Ed.* 2004, *39*, 2334-2337.
[188] Zhang, J.-P.; Lin, Y.-Y.; Zhang, W.-X.; Chen, Z.-M. *J. Am. Chem. Soc.* 2005, *127*, 14162-14165.
[189] Toh, N. L.; Nagarathinam, M.; Vittal, J. *Angew. Chem. Int. Ed.* 2005, *44*, 2237-2239.
[190] Chen, C.-L.; Goforth, A. M.; Smithm, M. D.; Su, C.-Y.; Zur Loye, H.-C. *Angew. Chem. Int. Ed.* 2005, *44*, 6637-6639.
[191] Dalgarno, S. J.; Thallapally, P. K.; Barbour, L. J.; Atwood, J. L. *Chem. Soc. Rev.* 2007, *36*, 236-238.

[192] Eddaoudi, M; Kim, J.; Rosi, N.; Vodak, D.; Wachter, J.; O'Keeffe, M.; Yaghi, O. M. *Science* 2002, *295*, 469-472.
[193] Kitagawa, S.; Kondo, M.; Seki, K. *Angew. Chem. Int. Ed.* 2000, *39*, 2082-2085.
[194] Zhao, X.; Xiao, B.; Fletcher, J.; Thomas, K. M.; Bradshaw, D.; Rosseinsky, M. J. *Science* 2004, *306*, 1012-1014.
[195] Barbour, L. J. *Chem. Commun.* 2006, 1163-1164.
[196] Atwood, J. L.; Barbour, L. J.; Jerga, A. J. *Angew. Chem. Int. Ed.* 2004, *43*, 2948-2950.
[197] Thallapally, P. K.; Dalgarno, S. J.; Atwood, J. L. *J. Am. Chem. Soc.* 2006, *128*, 15060-15063.
[198] Atwood, J. L.; Barbour, L. J.; Thallapally, P. K.; Wirsig, T. B. *Chem. Commun.* 2005, 51-53.
[199] Thallapally, P. K.; Wirsig, T. B.; Barbour, L. J.; Atwood, J. L. *Chem. Commun.* 2005, 4420-4422.
[200] Thallapally, P. K.; Lloyd, G. O.; Barbour, L. J.; Atwood, J. L. *Angew. Chem. Int. Ed.* 2005, *44*, 3848-3851.
[201] Thallapally, P. K.; Kirby, K. A.; Atwood, J. L. *New J. Chem.* 2007, *31*, 628-632.
[202] Thallapally, P. K.; McGrail, B. P.; Dalgarno, S. J.; Schaef, H. T.; Tian, J.; Atwood, J. L. *Nat. Mater.* 2008, *7*, 146-151.
[203] Thallapally, P. K.; McGrail, P. B.; Dalgarno, S. J.; Atwood, J. L. *Cryst. Growth Des.* 2008, *8(7)*, 2090-2092.
[204] Thallapally, P. K.; Dobrzanska, L.; Gingrich, T. R.; Wirsig, T. B.; Barbour, L. J.; Atwood, J. L. *Angew. Chem. Int. Ed.* 2006, *45*, 6506-6509.
[205] Thallapally, P. K.; McGrail, B. P.; Atwood, J. L. *Chem. Commun.* 2007, 1521-1522.
[206] Thallapally, P. K.; McGrail, B. P.; Atwood, J. L.; Gaeta, C.; Tedesco, C.; Neri, P *Chem. Mater.* 2007, *19*, 3356-3360.
[207] Shatz, J.; Schildbach, F.; Lentz, A.; Rastatter, S. *J. Chem. Soc. Perkin Trans. 2* 1998, 75-81.
[208] Benevellim, F.; Kolodziejski, W.; Wozniak, K.; Klinoski, J. *Phys. Chem. Chem. Phys.* 2001, *3*, 1762-1766.

K

kinetic parameters, 37
kinetics, 19

L

Langmuir, 4, 24, 35, 68
Langmuir-Blodgett, 35
lattice, 3, 45, 49
lattices, 45
ligand, 17, 21, 27, 32, 33
ligands, 10, 11, 18, 32, 33
linear, 39
liquid phase, 10
liquids, xiv, 39
lithium, 11
location, 20, 57
London, 64
low temperatures, 38, 56

M

magnetic, viii
mass spectrometry, 2, 16, 20, 47
matrix, 5, 19, 36
maximum water sorption, 43
measurement, 24
media, 6, 10
melamine, 18
Mendeleev, 66
metal ions, 11, 16, 47
methanol, 6
methylene, 6, 47
methylene chloride, 6
mobility, 45
model system, 55
modeling, 5
moieties, 18, 31, 47, 55
molar ratio, 57, 58
molar ratios, 57
molecular dynamics, 30
molecular structure, xiii, 12, 39
molecular weight, 24

molecules, xiii, 1, 3, 4, 5, 6, 7, 9, 10, 15, 18, 20, 21, 24, 27, 28, 36, 37, 43, 45, 47, 48, 51, 52, 53, 54, 55, 56, 58
monomeric, 18
monomers, 18
motion, 46
movement, 31

N

nanotechnology, 48
naphthalene, 24
network, xiv, 3, 40, 45
New York, vii, ix
nitrobenzene, 9
NMR, 9, 21, 22, 24, 48

O

observations, 58
oil, 44
oligomeric, 20
organ, 16
organic, xiv, 1, 2, 3, 4, 6, 7, 15, 18, 21, 27, 35, 36, 37, 39, 41, 47, 48, 55
organic compounds, 1, 35, 39
organic solvent, 4, 37, 47
organic solvents, 4
organometallic, iii, 16
oxygen, 2, 17, 28, 30, 31, 43

P

pairing, 16
parameter, 40
particles, 5
penalty, 56
phase diagram, 44
phase transformation, 51, 55
phase transitions, xiii, 3, 55
phenol, 31
phenolic, 15, 31
plasma, 20
play, 1

Index

ethers, 1, 11, 16, 17
ethyl acetate, 6
evaporation, 22, 30, 37
exposure, 38, 48
extrapolation, 30

F

failure, 10, 58
film, 35, 36, 44, 45
films, 4, 35, 36, 37, 44
flexibility, 2, 6, 15, 17, 21, 27, 30, 31, 32, 33, 40
flushing, 36
Fourier, 19
free energy, 2, 11, 33, 40, 47
free volume, 52
FTICR, 19, 34

G

gas, 2, 15, 16, 17, 19, 20, 21, 24, 27, 28, 29, 30, 31, 32, 33, 34, 40, 41, 44, 48, 56, 57, 58
gas chromatograph, 40, 41
gas phase, 2, 20, 21, 27, 28, 29, 30, 31, 32, 33
gas sorption, 56, 57
gases, xiv, 49, 56, 58
Gibbs, 21, 29, 30, 33
Gibbs energy, 21, 33
glass, 23
gold, 37
gravimetric analysis, 48
groups, 4, 6, 7, 11, 26, 32, 37, 45, 52, 55, 58

H

halogenated, 49
H-bonding, 40
heating, xiv, 53
hexane, 7, 36
high pressure, 57

Honda, 60
host, xiii, 2, 3, 4, 6, 9, 10, 15, 16, 17, 18, 20, 21, 27, 32, 33, 34, 36, 39, 40, 41, 45, 46, 47, 51, 52, 53, 55, 57, 58
humidity, 41
hydro, 35
hydrocarbons, 36
hydrodynamics, 5
hydrogen, xiii, 18, 31
hydrogen bonds, 31
hydrophobic, 5, 45, 46
hydroxyl, 15, 18
hypothesis, 7
hysteresis, 43

I

impurities, 24
in situ, 36
inclusion, 1, 3, 6, 7, 19, 21, 39, 40, 41, 47, 48, 51, 52, 53, 55, 57, 58
inclusion parameters, 39, 40
infrared, 48
infrared spectroscopy, 48
injury, viii
insight, 55
Inspection, 31
instruments, 19, 34
interaction, 1, 3, 15, 16, 17, 27, 34, 37, 39, 41
interactions, 1, 3, 4, 7, 10, 15, 16, 17, 23, 30, 34, 36, 38, 50, 52, 55
intercalation, xiii, 4, 45
interface, 49
interference, 34
intermolecular, 23
intermolecular interactions, 23
intrinsic, 3, 16, 34, 49
inversion, 10
ionic, 10
ionization, 2, 16, 18, 20
ions, 10, 15, 16, 17, 19, 47
isothermal, 44
isotherms, 40, 41, 42, 44

carbamic acid, 11
carbon, iv, 5, 20, 48, 56, 58
carbon dioxide, 48, 56
cation, 1, 2
cavities, xiv, 3, 4, 27, 36, 38, 45, 48, 49, 54, 55
cell, 23
Cellulose, vi
CH_4, 48, 58
channels, 5, 45
chemical reactions, 19
chiral, 16
chloride, 6
chlorobenzene, 53
chloroform, 6, 9, 36, 37
clean air, 36
closure, 18
clusters, 5
CO_2, iv, 48, 56, 57
coatings, 35
communication, 56
competition, 10
complement, 17, 27
complementarity, 39
components, 1, 10, 40
composition, 4, 21, 26, 28, 39
compounds, xiii, 1, 3, 4, 9, 11, 22, 25, 26, 39, 51, 53, 56
concentration, 36, 41
condensation, 24, 36
conductivity, 35
configuration, 40
conformational analysis, 31
Congress, viii
constraints, 3, 16, 17, 27, 34
construction, 47
conversion, 57
correlation, 30
covalent, 16
crown, 1, 7, 16, 17, 21, 27, 32
crystal lattice, 4, 27, 45
crystal structure, 1, 3, 5, 7, 52
crystal structures, 2, 7
crystalline, xiv, 4, 5, 43, 47, 49
crystallization, 22

crystals, 37, 45, 46, 47, 49, 52, 55, 58
cyclodextrins, 10, 16
cyclotron, 19

D

DCI, 20
decomposition, 24, 40, 41, 52
decomposition temperature, 40, 41
deformation, 5, 45
density, 56
deposition, 1, 4, 35
derivatives, xiii, 11, 16, 36, 56
desorption, 20, 37, 42
detection, 1, 17, 35
diffraction, 52, 58
diffusion, 37, 46
diffusion process, 38
dimer, 55
dimeric, 3, 51, 54
dimethylformamide, 11
disorder, 7
DNA, 16
donor, 18, 39

E

effusion, 4, 21, 23, 24
electrodes, 37
electron, 20, 47
electrons, 17, 28
electrostatic interactions, 30
encapsulated, 49
encapsulation, 49
energy, 12, 17, 20, 30, 31, 33, 34, 39, 56, 58
entrapment, 49
entropy, 11, 25, 29, 30, 31, 32, 33
environment, 1, 35
environmental control, 37
equilibrium, 10, 37, 41, 57
ESI, 16, 18
esters, 10, 19
ethanol, 12

INDEX

A

absorption, 4, 37, 56
acceptor, 18, 39
accounting, 51, 52
accuracy, 24, 42
acetate, 6
acetone, 6, 9, 58
acetonitrile, 7, 9, 11
acid, 6, 12, 18, 24, 41
acidic, 7
acidity, 7
acoustic, 35
activated carbon, 5
activation, 37
adducts, xiii, 3, 53
adsorption, 3, 4, 35, 36, 44, 56
aggregates, 16, 18, 19, 26
air, 36, 42, 49
alcohol, 6
alcohols, 11, 19
alkali, 10, 16, 47
alkanes, 49
alternative, 50
alters, 10
amines, 9
amplitude, 31
aniline, 36
aqueous solution, 11, 41

aqueous solutions, 11, 41
aromatic rings, 17, 28, 37, 45
assessment, 48
atoms, 2, 20, 30, 31, 43

B

barrier, 56
battery, 24
benzene, 6, 9, 20, 35, 36, 37
binding, xiii, xiv, 2, 3, 4, 10, 16, 17, 27, 28, 29, 34, 38, 39, 47, 49
binding energies, 38
binding energy, 40
biological systems, 1, 5
boiling, 40, 49
bonding, 18, 31
bonds, 31, 45
Boron, v

C

Ca^{2+}, 12
calixarenes, xiii, 1, 3, 5, 10, 16, 17, 19, 27, 28, 30, 37, 41, 42, 43, 44, 45, 46, 49, 52, 54
candidates, 48
capacitance, 35
capillary, 36
capsule, 51

polar groups, 11
polarity, 11
polarization, 48
polycrystalline, xiv
polyether, 2, 27, 32
polymer, 35
polymers, 48
polymorphism, 3
pores, 5, 46
porosity, 43, 47, 48
porous, 5, 49, 55
porous materials, 55
potassium, 47
powder, xiv, 52, 57, 58
powders, 41, 58
prediction, 39
pressure, xiii, 4, 24, 25, 37, 41, 44, 52, 57
propane, 56
protein, 16
purification, 49

Q

quadrupole, 19
quartz, 35, 37

R

range, xiii, 6, 21, 56
reactivity, 19
reagent, 19
reagents, 19
receptors, xiii, 1, 3, 15, 17, 21
recognition, 1, 6, 16, 37, 39
recovery, 36
recrystallization, xiii
recrystallized, 3
refractive index, 36
relationship, 2, 39
relationships, 40
relevance, 37
retention, 55
Reynolds, 68

rigidity, 6, 40
room temperature, 56, 58
rotations, 45
Royal Society, 60

S

salt, 47
salts, 18
sample, 23
scarcity, 16
search, 48
selectivity, xiii, 2, 3, 6, 9, 15, 17, 20, 27, 39, 47
self-assembly, 4, 35
sensitivity, 40
sensors, 1, 35, 37
separation, 55, 58
series, 18, 41, 53
services, viii
shape, 5, 40
signs, 29
simulation, 47
simulations, 30
single crystals, 3, 45, 52
single-crystalline, xiv
sites, 2, 17, 27, 38, 43, 45
solid phase, 52
solid state, 5, 6, 9, 30, 33, 45, 47, 49, 55
solid-state, xiv, 48
solubility, 11, 15
solvation, 3, 12, 15, 39
solvent, 3, 4, 7, 9, 10, 15, 17, 21, 22, 24, 27, 28, 29, 34, 37, 39, 48, 56, 58
solvent molecules, 4, 9, 21, 24, 27, 28, 37, 47, 56, 58
solvents, 4, 11, 30, 37, 56, 58
sorbents, 48
sorption, 5, 40, 41, 42, 44, 45, 48, 56, 57
sorption isotherms, 40, 41, 42, 43
species, 10, 11, 18, 19, 20, 51
spectroscopy, 2, 21, 37, 48
spectrum, 19
spin, 35
SPR, 36

stability, 1, 2, 10, 11, 19, 49, 55
stabilize, 4, 27
stages, 28
stainless steel, 23
steel, 23
steric, 17, 34
stoichiometry, xiv, 3, 18, 29, 39, 40, 41
storage, 41, 48, 55
strength, 2, 34
strong interaction, 17, 27
substrates, 37
sulfur, 47
sulfuric acid, 41
supramolecular, 1, 3, 16, 17, 27, 34, 38, 39
supramolecular chemistry, 16
surface area, 24, 44, 45
surface layer, 44
swelling, 36
symmetry, 7, 40
synthesis, 1, 11

T

temperature, 4, 20, 21, 24, 26, 28, 30, 31, 37, 40, 42, 51, 56, 58
temperature dependence, 4, 21, 26, 28, 30
temperature gradient, 24
thermal evaporation, 37
thermal stability, 1
thermodynamic, xiii, 3, 10, 16, 21, 28, 29, 32, 33, 34, 37, 40, 41, 49, 56
thermodynamic cycle, 28
thermodynamic function, 12, 28, 29, 32, 33
thermodynamic parameters, 11, 21, 34, 41
thermodynamic properties, 3, 16
thermodynamic stability, 49
thermodynamics, 10
thermogravimetric, 48
thermogravimetric analysis, 48
thermogravimetry, 41
thin film, 1, 4

thin films, 4
three-dimensional, 17, 27, 45
threshold, xiii, 40
time-frame, 19
toluene, 6, 9, 35, 36, 37, 51, 52
trajectory, 31
transduction, 35
transfer, 21, 28, 29, 30, 31, 32, 33
transformation, 50, 51, 55, 57, 58
transformations, 55
transistors, 35
transition, 41, 47, 51, 53
transitions, 31, 52
transport, 5, 46, 48
transportation, 56
two-dimensional, 44

U

urea, 18

V

vacuum, 24, 37, 41
validity, 2, 19
values, xiii, 24, 27, 29, 30, 33
van der Waals, 5, 52, 55
vapor, 4, 16, 21, 24, 25, 26, 35, 36, 40, 41, 44, 52, 55
variation, 39
voids, 49
volatilization, 20

W

water, 5, 10, 11, 41, 42, 44, 45
water absorption, 45
water clusters, 5
water diffusion, 46
water sorption, 5, 41, 42, 44
weak interaction, 17
weight loss, 24
workers, 49

X

X-ray analysis, 58
X-ray crystallography, 47
X-ray diffraction, 12, 45, 48, 57
X-ray diffraction data, 45
xylene, 6, 9, 36

Y

yield, 53

Z

zeolites, 3, 5